Great Scientists Speak Again

Great
Scientists
Speak
Again

by
Richard M. Eakin

University of California Press
Berkeley · Los Angeles · London

This book is dedicated to thousands and to one. The many are my students, not only those in the University of California course Zoology 10, for whom these lectures were first prepared, but also those in embryology and my graduate students. The one is my wife, Mary, to whom I am indebted for forty years, as of this date, of devotion and encouragement.

University of California Press
Berkeley and Los Angeles, California

University of California Press, Ltd.
London, England

Copyright © 1975, by
The Regents of the University of California
ISBN: 9780520306295
Second Printing 1982

Library of Congress Catalog Card Number: 74-22960

2 3 4 5 6 7 8 9

Contents

Acknowledgments

I AM GRATEFUL for an award toward the cost of the figures from an institutional grant by the National Science Foundation. I acknowledge also some assistance through a grant, GM 10292, from the United States Public Health Service. Four of the lectures were published in preliminary form and without illustrations in *Bios* in 1970–1971. I appreciate the permission by that journal to use those publications as the basis for chapters 2, 3, 4, and 6.

It is a pleasure to acknowledge Alfred Blaker and Gene Groppetti of the Scientific Photographic Laboratory on the Berkeley campus for the photographs of the characters; Emily Reid, scientific illustrator of my department, for the diagrams and sketches; Roberta Vance and Shan Otey of the costume shop of our Department of Dramatic Art for counsel on and construction of costumes; Robert Buckles, make-up artist (I now do my own make-ups); Nancy Jarvis and Henry Bennett of the University of California Press, for editing and designing the book; Teriann Asami for the ribbon copy of the manuscript; and Carol Reed, John Gayman, and Madeleine Bergman for general assistance. I thank also the following persons for support, advice, and encouragement: Dean Sanford S. Elberg; professors Max Alfert, Fred Harris, Ralph I. Smith, Curt Stern, and Marvalee Wake; Kenneth C. Owen, business administrator of my department; Jean Brandenburger, my research associate; and my wife, Mary Mulford Eakin. Other acknowledgments are made in the prefaces to the individual lectures.

R.M.E.

Addendum at second printing.
I am pleased with the favorable reception of *Great Scientists Speak Again*. Both cloth and paper back editions have sold out, the former several years ago. In reprinting the book I have the opportunity to correct some errors, to acknowledge the source of figure 51, and to call the attention of the reader to three recent books, among many, on Darwin and his theory of evolution (see page 119). I wish that it were within my financial resources to replace all of the "Charles Darwin" photographs with ones showing my new makeup with hairpieces that give me a closer resemblance to CD.

R.M.E.

Introduction

IN THE LATE sixties students at Berkeley made clear to the faculty, in various ways, that it was time to reconsider old patterns of instruction and to introduce new ones. Lectures particularly came under fire. I observed increasing absenteeism and inattention in my lectures in general animal biology, listed as Zoology 10 in the Berkeley catalogue. What could I do to recapture the students' interest in and enthusiasm for biology? Of course, there were many, as always, who find any subject exciting despite a dull lecturer.

One morning in the shower I was struck not only by the spray but by an idea: dress up and make up as some of the great biologists, and present their discoveries, thoughts, and philosophies in their own words. Portray them as persons with hopes and ambitions, with frustrations over failures and joy from success, with their strengths and human weaknesses. Then, on second thought, fear that I might be laughed out of the auditorium seized me. I had no schooling in dramatic art.

1

But—as some friends tell me—there appears to be considerable "ham" in my nature.

With the assistance of Professor Emeritus Fred Harris of our Department of Dramatic Art, I secured costumes and the services of a skilled make-up artist, Robert Buckles. And one day I appeared before my large class in elementary biology as William Harvey, unannounced, in a wig and Elizabethan dress, and with some props including a heart and a pitcher of tomato juice (blood). "Harvey" discussed his great discovery of the circulation of blood and demonstrated his brilliant use of observation, experimentation, calculation, and inductive reasoning to refute the ideas of Galen, the Greek physician to the Roman emperor Marcus Aurelius—ideas that had been almost law for fourteen centuries. "Harvey" concluded by admonishing his audience, to wit: "Listen to your instructors with respect and read carefully that which is written, but do not hesitate to believe your own senses." The lecture was a success.

I went on to present four other impersonations: Gregor Mendel, Charles Darwin, William Beaumont (who studied gastric digestion in the stomach of a fur-trapper, Alexis St. Martin), and Hans Spemann (1935 Nobel Laureate in Physiology and Medicine, who discovered the organizer principle in embryonic development). And this year, 1974, I added the sixth, and probably the last, character: Louis Pasteur.

I use these "guest" lecturers to introduce different segments of the course. The order of appearance here is that in Zoology 10. Then I follow each one with two to four lectures on the subjects as they are presently understood. For example, "Mendel" introduces the topic of genetics, and I then bring the historical account to modern times and conclude with a lecture on human genetic diseases. In a quarter term of thirty lectures, the "guest" speakers appear approximately fortnightly. To my gratification my own lectures are better attended, and I receive

closer attention because of the excitement in the various subjects initially generated by the "guest" scientists.

Admittedly, these impersonations are an unusual solution to the problem of making lectures more interesting to
students, and I recognize that although any teacher is to some
extent an actor on a stage (particularly in a large lecture course)
such a solution is not suited to every instructor's inclination or
ability. Perhaps my experience may stimulate others to find
different but equally rewarding ways of presenting their subject
in a manner that captures the attention and interest of this
generation of students, who are, I regret to say, addicts of the
"boob-tube" (which for any foreign reader is American slang
for television).

Following some worldwide publicity (*Life* magazine, *International Herald Tribune, Der Spiegel*, etc.), I received invita-
tions to deliver the lectures from Florida to Newfoundland. I
am able to accept only a few each year. To meet the demand,
four of the impersonations have been placed on film for use in
classrooms, museums, and academies, and by scientific organi-
zations. They are available by rental or purchase through the
Media Center of the University of California, Berkeley. But the
movie versions are much abridged, and the number of students
who can be reached through this medium is limited. It was
suggested that the full lectures be published in book form. To
that end they have been revised and illustrated with some of
the sketches and slides used by the "guest" lecturers and by
photographs of the six speakers taken mostly at the time of the
1974 presentation to Zoology 10.

W.H.

William Harvey

1578–1657

WILLIAM HARVEY is one of my heroes. A portrait of
him hangs by my desk, reminding me daily of his rigorous
scientific methods. My resemblance to him in the accompany-
ing photographs is not so good as I would like. The wig, a
beautiful and expensive French creation, is not quite right, and
I am much too tall to play "little Doctor Harvey," as he was
known to his friends. I have been guided by biographies of
Harvey and, of course, by his great book, *De Motu Cordis.*
Unfortunately, my high-school Latin is inadequate to read the
book in the original. Harvey would say, "Oh, what a pity." I
like Chauncey Leake's translation for its directness and because,
as he says, Harvey was "terse and snappy." For example, in
describing his search for Galen's pores in the septum of the
heart Harvey says, according to the English translator Robert
Willis: "But in faith, no such pores can be demonstrated."
Leake's translation is, "Damn it, no such pores exist."

I have been assisted also by a remarkable film produced in
black and white by the Royal College of Physicians, London,

on the occasion of the tricentennial of the publication of
Harvey's book (1628) and in color on the three-hundredth
anniversary of the death of Harvey (1657). This film reproduces Harvey's stunning experiments and demonstrations on
the motions of the heart and the circulation of blood, with
commentary in Harvey's own words which I have generously
borrowed.

The assistant upon whom "Harvey" demonstrates the
valves in the veins (see fig. 8) was Jeffrey Baylis, my associate in
Zoology 10 in 1974. Jeff was also "Pasteur's" assistant in that
lecture as told here, and he played these two roles in the motion
picture impersonations of Harvey and Pasteur. For the last
photograph in this chapter I am indebted to Dr. Perry Turner,
an assistant in Zoology 10 in 1971.

The Harvey Lecture

[William Harvey, a vigorous physician in his late sixties, enters
the auditorium in Elizabethan dress of knee britches, embroi-
dered doublet with short cape and broad linen neckpiece, white
stockings, and shoes with large buckles. His long flowing
hair, two-piece beard, and moustache are light grey. He carries
a book, his *Exercitatio Anatomica de Motu Cordis et Sanguinis
in Animalibus*. He moves to center stage with short, quick
steps, places the book on the lecture table, and raps for attention on the podium.]

Gracious ladies and worthy gentlemen. Before presenting
a discourse on the heart and the circulation of blood, listen, I
pray you, to some personal history. I was born on All Fool's
Day, 1578, at Folkestone on the coast of Kent, the eldest of a
week of sons of Thomas Harvey, a respected official of the

township. I was bred to learning, but my six brethren, alas poor fellows, were bound as apprentices in London. I entered Gonville and Caius College, Cambridge, with the intent of studying medicine and completed the bachelor's degree four years later, at the age of 19. This was in 1597, yes, 1597—the same year that Sir Walter Raleigh made his gallant expedition against the bloody Spaniards.

The following year I journeyed to Italy (where so much was stirring in the arts and sciences) to that renowned center of learning, Padua, and to the *Universitas juristarum*, the scene of triumph of Vesalius, Renaldus Columbus, Galileo, and my teacher of anatomy, Hieronymus Fabricius of Aquapendente. A new amphitheater had been recently erected for his anatomical demonstrations. Methinks, I was a favorite pupil of the honorable Fabricius because I was appointed to assist him in certain exercises. Whilst in Padua I performed many dissections on serpents and frogs and fishes which I obtained from the marketplace. Sometimes the fishes were still flipping and flopping, and I had the good fortune to observe the last struggling beats of their hearts. I did often reflect upon the works of that great authority, Galen, the celebrated Greek physician to the Roman emperor Marcus Aurelius. Galen's views had been accepted as law for fourteen centuries.

According to Galen the blood is formed in the intestines, and then in the liver it is charged with *natural spirits* and carried to the right side of the heart by the veins. [Harvey uses a diagram on the chalkboard, fig. 1.] Contractions of right auricle and ventricle cause the blood with natural spirits to ebb and flow through the veins to all parts of the body. Some of the blood was supposed to ooze by invisible pores in the septum of the heart into the left ventricle, there to be mixed with a certain quantity of air, drawn from the lungs through the pulmonary veins, and stuffed with *vital spirits*. Pulsations of the left ventricle then move the vital spirits back and forth in the

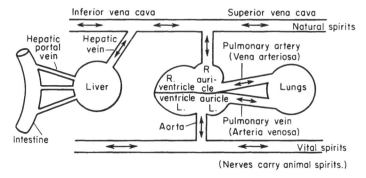

[Galen called the pulmonary artery a vein because it carried natural spirits, but as it was like an artery structurally he used _vena arteriosa_. And the pulmonary vein was termed an artery because it transported vital spirits, but as it resembled a vein it was designated _arteria venosa_.]

FIG. 1.   *Flow of blood through the body according to Galen.*

arteries to heat and quicken and restore all tissues of the body. Some of the vital spirits upon reaching the brain were supposed to be changed into *animal spirits* which are distributed by hollow nerves to the muscles to effect movements.

By my young beard I did not believe these notions, but I gave the proper answers to the jury, you may be sure; and in 1602 I was awarded the diploma Doctor of Physic. I straightway returned to England, just in time to attend a new play by Shakespeare entitled *Hamlet.* Two years later there happened three events of note: there was a ghastly plague in London, I became a candidate of the Royal College of Physicians, and I married the lovely Elizabeth Browne. I might add that in wooing this fair maiden I was not unmindful–ah–[Harvey takes a pinch of snuff.] Ah, I was not unmindful that her father, Dr. Lancelot Browne, was physician to the king, James the First. After the death of the good doctor–alas, poor fellow–I became physician to his Royal Highness. And, Lord Chancellor Francis Bacon was also a patient of mine. In 1607 I was elected fellow of the Royal College of Physicians–the same year that we established our colony at Jamestown in Virginia. In 1615 I was appointed Lumleian Lecturer in Anatomy at the Royal College, and in my first visceral lecture I made so bold as to announce some of my heretical views regarding the circulation of blood. Incidently, I usually lecture in Latin, but I was given to understand that you know no Latin–oh, what a pity–so I am giving this discourse in English. Meanwhile, I carried out dissections on the bodies of the newly dead at St. Bartholomew's Hospital, in utter secrecy, of course, because of the existing legislation. And I continued to perform experiments on animals in my laboratory. I was a member of many of King James's royal hunting parties. Later I was physcian also to Charles the First, he who recently lost his head. [Harvey draws a forefinger across his throat.] On these sorties I had opportunities to observe the flagging hearts of the game and to test some of my theories. I finally composed my arguments in this book, *Exercitatio Anatomica de Motu Cordis et Sanguinis in Animalibus,* which was published in 1628 by Wilhelm Fitzer of Frankfurt. It appeared just in time for the great book fair in that city.

My views on the heart and circulation rest not on books nor the opinions of my predecessors (fig. 2), but on direct, ocular observations, dissections, and experimentation. They may be summarized as follows. [Harvey uses another diagram on the chalkboard, fig. 3.] The blood returns from the tissues of the body to the heart by way of the large veins: the superior vena cava from the head and the inferior cava from the lower parts. Together they enter the right auricle. Thence, the blood moves through the tricuspid valves, which open like flood gates, into the right ventricle. A contraction (systole) of the ventricle discharges the blood past three semilunar valves into the pulmonary artery and on to the lungs where noxious vapours are removed. The pulmonary veins return the purified blood to the left auricle, which communicates with the left ventricle by the bicuspid or mitral valves. Contractions of the left ventricle, beating simultaneously with the right one, hurl its contents past another set of semilunar valves into the great aorta which feeds the arteries of the body. The blood then percolates

FIG. 6.   *Damn it, no such pores exist.*

FIG. 7.   *The flow of blood to and from the hind limb of an animal.*

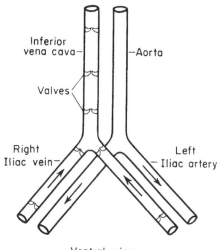

Inferior vena cava—

—Aorta

Valves

Right Iliac vein—

Left Iliac artery

Ventral view

be tolerated. Damn it, they do not exist. [Harvey pounds the podium, fig. 6.] Why, I ask you, must we imagine secret pores to carry the blood from right to left ventricles when there is so large and so open a pathway through the pulmonary arteries and veins and the soft tissues of the lungs? I have often reflected on the symmetry of the heart and on the size of the vessels entering and leaving it. Why would Nature, making nothing in vain, give these vessels such greatness uselessly? On this and other questions I have long pondered, and deeply.

I have asked myself: how much blood is transmitted by each beat of the heart? Let us measure it. [Harvey picks up a human heart from which the auricles have been removed.] Fill a ventricle, say the left one, to capacity. [Harvey pours water from a small pewter pitcher into the left ventricle to over-flowing and then empties it into a measuring vessel.] I have found a ventricle to hold upwards of two ounces. [Harvey writes on the chalkboard: ℥ ii.] Now let us suppose that only one fourth of that amount is expelled at each pulsation of the ventricle. That would be four drachms. [Harvey writes: ℨ iv.] If a heart beats a thousand times—taking about fifteen minutes in a normal subject—it will have pumped 4,000 drachms. [Harvey's notation is ℨ MMMM.] Why, that is a greater quantity of blood than all the blood in the body. Whence came that blood? Whither goes all that blood? It is not possible for the digested aliment to furnish such an abundance without the heart draining the veins and rupturing the arteries, unless—as I began to think—there is a sort of motion, as it were, in a circle with the same blood passing through the heart again and again.

I have yet to consider the veins and arteries to a part of the body, such as a limb. Expose, for example, the vein and artery supplying the leg of an animal. By dissection we learn that the iliac artery is a branch of the aorta and the iliac vein a tributary of the inferior vena cava. [Harvey refers to a third diagram on

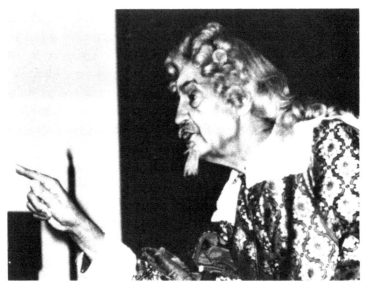

FIG. 9.   *But do not hesitate to believe your own senses.*

moved in the reverse direction because of the valves. [Harvey removes the bandage from the arm of Baylis.] I thank thee for thy good services. Go in peace.

Thus, by ocular observations and dissections, by experiments and measurements, and by calculations and inductive reasoning, it is absolutely necessary to conclude that the blood is propelled by the heart in a circle and that this is the only end of the heart. And now in parting, this admonition: listen to your teachers with respect, and carefully read that which is written, but do not hesitate to believe your own senses [fig. 9].

Gracious ladies and worthy gentlemen, I bid you good day.

[Harvey nods to his audience, picks up his book, and quickly leaves the auditorium.]

W. B.

William Beaumont

1785–1853

A QUESTION frequently asked of me is, "Which is your favorite character?" The answer varies, depending upon the role last enacted. I play two of my characters humorously: Mendel and Beaumont. The former is a jolly priest with secular tastes to flavor his biology, and the latter a rough-and-ready army surgeon who is always quick to fight on any question of intellectual integrity or personal honor. My interpretation of Beaumont has been guided by his book, *Gastric Juice and Physiology of Digestion*, a classic in human physiology, and by *The Life and Letters of Dr. William Beaumont* by Jesse S. Myer (1912). Helpful also was a biographical essay by Sir William Osler entitled "A Pioneer American Physiologist," which was first delivered as an address before the St. Louis Medical Society on October 4, 1902, and published with a facsimile edition of Beaumont's book in 1929 and reprinted in 1941. Some of the colloquialisms I place in the speech of Beaumont, such as *skidaddle* and *spizzerinktum*, were learned from my grandfathers. I have never seen a portrait or photograph of Beaumont with a

FIG. 11.   *The Chandlers had a very attractive daughter.*

FIG. 12.   *Beaumont's chalkboard diagrams.*

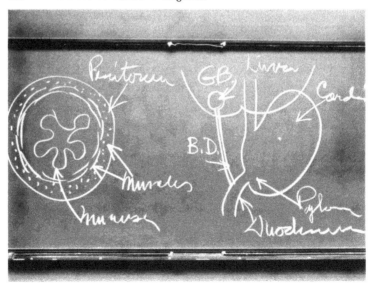

writin' and 'rithmetic, which I did for three years after leaving the family farm in Connecticut, with a horse and cutter, a barrel of cider, and some hard-earned cash. Farming was not to my liking either. You see, my father—but no need to go into that. On with the anatomy of the stomach.

The stomach is a big bag on the left side under the floatin' ribs. Anatomy books show it something like this. [Beaumont draws a crude sketch on the chalkboard, fig. 12.] It looks like a bagpipe. I reckon they used stomachs for the first bagpipes. Alexis's stomach, however, was always changing its shape. Oh, I'll come to Alexis presently. The gullet or esophagus enters here, discharging the *materia alimentaria* into the stomach. The superior end of the organ is a large bay, and because it lies near the heart it is called the cardiac region. [Beaumont scrawls the term on the board.] At the opposite end, the stomach narrows. This is the pyloric region, and here the aliment, now changed into chyme, passes by way of the pyloric valve into the first part of the small intestine, called the duodenum. On the right side of the abdomen and overlapping the stomach is the liver, which weighs about four pounds. The liver, as you know, produces bile which collects in the gall bladder, and there it acquires its thickness and its bitterness before flowing down the bile duct to the duodenum. [Beaumont adds these structures to his sketch.]

Believe me, I have seen a large number of bellies split open—one of the advantages of being an army surgeon. [Beaumont reminisces again.] You see, I quit Doc Chandler, under whose friendly inspiration and instruction in physic and surgery I had happily studied for two years to my satisfaction and that of my preceptor. I could cup and bleed and amputate, and I was licensed to practice by the State of Vermont. But I had a hankering to join the army. Besides I was not making any progress in wooing Mary Chandler. The British were blowing up war clouds again; and here was a good chance to display my patriotism, to put my learning into immediate practice, and to earn some sorely needed cash. I was commissioned surgeon's mate by President Madison in 1812.

At the battle of York I had more than I bargained for. The British broke off the engagement, retreated, and then blew up the magazine. The explosion cut and mangled our men most shockingly. About 60 members of our regiment were killed and over 300 wounded. For forty-eight hours without food and sleep we surgeons cut and sawed and trepanned skulls. By the great horned spoon it was terrible. I was a war casualty, too —result of my own tomfoolery. Another surgeon's mate and I dared each other to stand close to a cannon when it was fired. I won, but I have been partially deaf ever since. And—[Beaumont becomes aware of his digression.] Well, back to the stomach.

This organ like others of the alimentary canal is made up of layers. Make a cross section of the stomach—like that. [Beaumont makes a line across his earlier sketch of the organ and then hurriedly draws a cross-sectional view, fig. 12.] It is covered on the outside with peritoneum, a glistening membrane; inside the peritoneum are two coats of muscle. In the outer one the fibers are arranged longitudinally, and when they contract the stomach shortens on its long axis. In the second and inner muscle layer the fibers are disposed in a circular fashion. Their contraction causes the diameter of the stomach to decrease. The motions of the stomach result from the coordinated action of these two muscle layers. Inside the muscle coats is a layer of connective tissue that binds the muscles to the inner lining or mucous coat. The last named has many glands that look like little bottles. They pour out mucus which coats the folded lining of the organ.

In Alexis—oh, I'll tell you all about Alexis in a moment—the folds of the stomach had a pinkish color, like the roses in Deb's garden in Saint Louis. Deborah's my wife. Yes, I live in Saint Louis now. I quit the army—didn't take to the new surgeon general because–well, no point in fighting that battle again. I have a lucrative practice in Saint Louis, and I have been offered the Chair in Surgery at the new medical school of Saint Louis University. Oh, I like Saint "Louie"–power in the air there.

Now, where in tarnation was I. Anatomy of the stomach–oh, well, I reckon that's enough anatomy to understand my discourse on the physiology of the organ. Before my humble studies on gastric digestion (thanks to Alexis), there was much speculation about the function of the stomach. Professor Dunglison's new textbook of physiology had restated several theories on how the stomach digests the aliment: by concoction, by putrefaction, by maceration, by trituration, and by fermentation. The learned Dunglison even quoted that pithy pronouncement of the celebrated British anatomist and physiologist, William Hunter, who addressed some physicians as follows: "Some physiologists will have it that the stomach is a mill, others that it is a fermenting vat, others again, that it is a stew pan; but in my view of the matter, gentlemen, it is neither a mill, nor a fermenting vat, nor a stew pan, but a stomach, gentlemen, a stomach!" Stomach indeed. I have often wondered what William Harvey would have said to that fatuous pronouncement. In the bright light of science and the deductions of experiment all of these theories have been prostrated in the dust. The process of developing truth by patient and perservering research is incompatible with these notions of unrestrained genius. The drudgery of science is all too often left to humbler and more unpretending contributors. These are, of course, general remarks, and I have no wish to apply them, so far as bad motives are inferred, to respectable gentlemen (who will go unnamed), who have been seduced by the ingenuity of their arguments or the blandishments of their style.

There have been, of course, good studies on gastric digestion. I refer especially to the work of the Italian abbot, Lazzaro

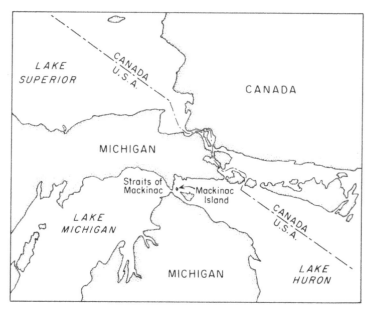

FIG. 13.   *Geographic position of Mackinac Island, Michigan Territory.*

Spallanzani. [Beaumont scribbles the name illegibly on the board.] He was a professor of physiology at several Italian universities. Spallanzani was a very skillful man and performed some clever experiments. For example, he took a small wire cage, stuffed it with meat, tied a string to it, and swallowed it. [Beaumont pantomimes.] He probably went about his priestly duties with the string dangling from the corner of his mouth. After a couple of hours he hauled up the cage–the meat was gone. [Further pantomime.] I have obtained identical results in experiments on Alexis. Yes, I am almost to Alexis.

Well, after the War of 1812 and a short period of private practice, I found that the army was in my blood, so I joined up again and was sent way out West to Fort Michillimackinac in Michigan Territory. The fort, called Mackinac for short, is on an island where the waters of Lake Huron meet those of Lake Michigan [fig. 13]. I was now Assistant Surgeon Beaumont and the only doctor within hundreds of miles of the fort. In summer the island was a bustling, boisterous community of trappers, traders, and voyageurs bringing their pelts to the post of the American Fur Company. Mark me, there were many fights and more opportunities to study anatomy and practice surgery.

On one beautiful day in June 1822, I was summoned to the company store because of an accident. A French-Canadian trapper by the name of Alexis St. Martin had been shot accidentally in the left side. The gun had discharged at close range, and the shot and wadding had torn open the poor fellow's rib basket, left lung, and stomach. The lad was only 19 years of age. When I arrived he was senseless and in a moribund state, and his breakfast was pouring out of the hole in his stomach. I fished out some of the shot, wadding, and pieces of clothing and covered the wound. I returned in an hour and to my surprise found him better than expected. I still thought that he could not live more than 36 hours. I dressed the wound more carefully, removed fragments of ribs, and pushed back the lung and stomach as much as possible.

I moved him to the fort hospital soon thereafter so that I could take better care of him. The wound began to heal and his fever to subside. There followed a series of carbuncles in which pieces of bone and cartilage were sloughed, but the hole in the stomach remained open despite all my efforts to promote union of its edges. To administer a cathartic of sulphur I used a method never before employed in the history of medicine: poured it directly into the stomach. [Further pantomime.] After many months Alexis was still in a helpless state. Finally, the county refused further assistance, declared Alexis a pauper, and told him to skidaddle home in an open batteau to Montreal, a distance of more than 1,000 miles. Repeated objections against such inhuman disposal of a person being of no avail and knowing that his life must be sacrificed, I resolved to rescue him from pending misery and death. I took him into my own family at a time when he was sick and suffering from the debilitating effects of his wounds, naked and destitute of everything but pain, a little breath of life, and a wounded body. After another year he began to show some spizzerinktum; he was able to walk and help himself a little. During all this time I nursed him, fed him, clothed him, lodged him, and dressed his wounds daily–sometimes twice daily.

Then, you know, I began to realize that I had a remarkable opportunity to study gastric digestion and that I had a duty to humanity to conduct observations and experiments. On removing the dressings I frequently found the stomach partly everted in the shape of a half-blown rose. I could easily reduce it by gentle pressure with not a trace of pain to the patient. When he lay on his right side I could look directly into the cavity of the stomach and almost see the process of digestion. I could put fluids and food into the organ with a funnel or spoon, and sometimes I introduced pieces of meat tied to a string. And I could remove gastric secretions or digested aliment with a siphon. I found I could extract a gill of gastric juice every two or three days and conduct experiments on digestion outside the body, comparing the process with that in the body. Various kinds of digestible and indigestible materials could be studied. By placing Alexis on his left side and by pressing on the liver [Beaumont pushes his right side vigorously], I could force bright yellow bile into the stomach through the pylorus and investigate this secretion.

Later, I learned how to get my own gastric juice. Simply swallow this tube, apply some suction with a bulb, and collect the fluid in a flask. [Beaumont pantomimes swallowing a stomach tube, fig. 14.] Well, I'll not put it all the way down; just had my lunch. But here is some of my gastric juice I collected this morning before breakfast. [Beaumont picks up a small beaker containing an acidic solution of pepsin.] Pure gastric juice is a clear, transparent fluid without odor, a little saltish, and very perceptibilibly acid. [Beaumont tastes the solution in the beaker with his finger.] Its taste is, when

applied to the tongue, similar to mucilaginous water acidulated with muriatic acid. It is diffusible in water, wine or
spirits; effervesces with alkalies; coagulates albumen to an eminent degree, and then dissolves it, as I shall now demonstrate.

To a test tube with coagulated albumen from a hen's egg
I add some of my gastric juice, and to a second tube, called
a control, I add only muriatic acid. Now let the tubes rest while
I continue this discourse.

FIG. 14.   *I learned to get my own gastric juice.*

Gastric juice is powerfully antiseptic, checking the putrefaction of meat and healing sores and foul ulcerating surfaces. Remarkable stuff. It commonly, perhaps always, contains mucus. My surmise that it has free muriatic acid was confirmed by analyses made by two chemists on samples which I sent to them: Professor Dunglison at the University of Virginia and Professor Silliman of Yale University. Incidentally, young whippersnapper chemists nowadays have changed the name of muriatic acid; they call it hydrochloric acid.

I communicated my first findings to Surgeon General Lovell, to whom your humble servant is indebted for much encouragement. Dr. Lovell replied favorably. And he sent my communication to the Philadelphia *Medical Recorder* for publication. Moreover, he suggested that I test the widely held opinion that the stomach digests foods in order of kind. All of one kind first, then all of a second, and so on. This was the supposed cause of the bad effects of a variety of foods in the same meal. Suppose a man eats beef, potatoes, cabbage, and pudding. Will the stomach digest the beef first, leaving the others untouched, then the pudding, then the potatoes, and the cabbage last? If the number of articles be a dozen, will the stomach handle eight or ten but then become exhausted leaving the remainder to ferment and cause indigestion? I made many experiments and found these notions to be nonsense, although it is true that some foods are more easily digested by the stomach than others, as I shall soon note.

Well, I began to feel my isolation on the frontier, and so I requested a transfer to the East where I could take advantage of better facilities for study and confer with professors of physiology and chemistry. In due time I was ordered to Fort Niagara, taking my family and Alexis with me. Soon thereafter Alexis took "French leave" and escaped into Canada. I made every effort to locate the truant, but without success. Although broken in spirit I summoned courage to give the world the

results of further observations by sending them to the *Medical
Recorder.*

Being blessed with ample determination, probably
handed down to me by my father, I persuaded the American
Fur Company to instruct its agents all through the West to be
on the lookout for my man. At last they found him, employed
him, and sent him to me, his benefactor, at great expense to
me. Alexis had married meanwhile, had two children, was in
robust health, but his stomach was essentially as I last saw it.
And I began experiments immediately.

I made a study of gastric temperatures by simply placing a
thermometer through the aperture, to correlate temperature
with the state of the stomach and the condition of the patient.
Alexis was a moody fellow, a heavy drinker, and at times very
ugly. But despite the difficulties engendered by these unfelici-
tous traits, they promoted the cause of science, because I
discovered that the mind, and alcohol, and even the weather
influenced gastric digestion. Some physiologists state that
there is a considerable increase of temperature of the stomach
during the digestion of a meal. Tommyrot. Active exercise,
however, will elevate the temperature about one and a half
degrees, whether the stomach is full or fasting.

Let me give you some good advice at this juncture. Keep
a full journal of your experiments and observations; let no fact
go unrecorded even though seemingly trivial at the time;
maintain an orderly journal; and record everything exactly as it
is. [Beaumont pounds the podium.] Have no expectations and
no theories to prove. I humbly submit an example from my
own journal.

August 7, 1825. 11 O'clock A.M. Weather partially cloudy; atmosphere
dry and smoky, wind E and light. Temperature, north exposure, 69 degrees
Fahrenheit. Temperature of stomach, 100 degrees and a fraction. Remained
stationary. Pulse 55 in recumbent position, 65 sitting erect. St. Martin had a

good night's sleep; was in pleasant mood; cooperative. Had fasted for 17 hours. Stomach healthy and completely empty. Introduced gum elastic tube through aperture and drew off one ounce of pure gastric juice unmixed with any other matter, except a trace of mucus. Placed juice in three ounce vial. Added piece of boiled beef weighing three drachms to vial, corked vial; placed it in a saucepan of water, raised temperature to 100 degrees and kept it at that level on a regulated sand bath. Immediately placed similar sized piece of beef of identical weight tied to string into stomach by aperture. Twelve noon, withdrew meat from stomach; looked same as that in vial. Meat soft and fluid opaque. One o'clock drew out string again; meat completely digested and gone. Cellular texture of piece in vial entirely destroyed, leaving some fibers loose and unconnected and floating about in fine shreds.

For about two years I conducted experiments on Alexis, all the while paying for his services, providing him and his family with subsistence, lodging, and other necessities. But his wife became very unhappy and homesick, and finally it was necessary to let them return to Canada, but only on the promise that Alexis would return for more experiments. As expected, he did not keep his promise. Within a year, however, he needed money, and I engaged him again, this time under a contract which we signed, Alexis by his mark. [Beaumont picks up a sheet of lecture notes.] This agreement read in part:

Said Alexis St. Martin binds himself for one year to serve, abide, and continue with said William Beaumont, wherever he shall go or travel or reside in any part of the world; that he will at all times during said term when thereto required by said William submit to assist and promote by all means in his power such philosophical or medical experiments as the said William shall direct or cause to be made on or in the stomach of him the said Alexis, either through and by means of the aperture or opening thereto in the side of him, the said Alexis.

To strengthen further my tenuous hold on the unreliable Alexis I persuaded the surgeon general to arrange an appointment in the army. And so Sergeant Alexis St. Martin was detailed to Surgeon William Beaumont to assist in the conduct of scientific research on gastric digestion. Eventually all of these

bonds dissolved like a piece of meat in Alexis's stomach, and I
have not been able to lure or coerce him into my service again.
He was, of course, the butt of many jokes: the man with a lid on
his stomach. And I suppose he became weary of my experiments. But I have not given up! [Beaumont shakes his fist.] I'll
get him back.

Meanwhile, it was time to make a final report to the
scientific world, which I prepared in the form of a book
detailing my 238 experiments precisely as I recorded them.
[Beaumont picks up his book.] It was published in 1833 at my
own expense, 280 pages, entitled *Experiments and Observations on
the Gastric Juice and the Physiology of Digestion.* To my gratification the book has been most favorably received in this country
and in Europe. Within a year after its publication a German
edition appeared. After describing my various observations and
experiments, I did allow myself some modest conclusions, 51 in
number! I have selected a few of the more important ones for
your edification. [Beaumont reads from his book.]

1. Animal aliments are more easy of digestion than vegetable.
2. Digestion is facilitated by minuteness of division, and tenderness of
   fibre, and retarded by opposite qualities.
3. Oily food is difficult of digestion by the stomach.
4. The use of ardent spirits [strong liquor] produces diseases of the
   stomach, if perservered in.
5. The action of gastric digestion is facilitated by the warmth and
   motions of the stomach.
6. Gastric juice contains free muriatic acid and other active chemical
   principles.

[Beaumont puts his book aside and turns to the experiments begun earlier in the lecture.] Now let us examine the
two tubes with coagulated albumen. You will note that the
tube with gastric juice is slightly clearer than the one with
muriatic acid alone. The difference between two similar tubes
prepared before the lecture is dramatic [fig. 15]. The coagulated

FIG. 15.   *Gastric juice has digested the albumen in the tube to your left.*

albumen is largely dissolved in the tube with acidulated gastric juice, whereas there is no change in the tube with muriatic acid alone. This experiment clearly demonstrates that there are certain chemical principles in gastric juice which digest the albumen. Muriatic acid alone will not dissolve the albumen or other foods.

To continue with more of my conclusions:

7. Exercise elevates the temperature of the stomach, whereas sleep or rest in a recumbent position depresses it.
8. Bile is not ordinarily found in the stomach, but it assists in the digestion of oily foods.
9. Water, ardent spirits, and most other fluids are not affected by gastric juice, but pass from the stomach soon after being received.
10. Bulk, as well as nutriment, is necessary to the articles of diet.

Enough of my conclusions. Now I am fully aware of the importance of the subject of gastric digestion, and I am willing to risk the censure or neglect of critics if I may be permitted to cast my mite into the treasury of knowledge and to be the means, either directly or indirectly, of subserving the cause of truth and improving the condition of suffering humanity. I have submitted here a body of facts that cannot be invalidated. Although my opinions may be doubted, denied, or approved, the facts are incontrovertible. Truth, like beauty, is most adorned when unadorned, and in prosecuting these experiments and inquiries I believe that I have been guided by its light.

And so gentlemen, ah—and ladies also, I thank you for your kind attention. [Beaumont quickly gathers up instruments, bandage, notes and book; stuffs them into his bag; and strides from the platform.]

Hans Spemann

1869–1941

OF THE SIX impersonations, my "Spemann" is surely the most authoritative, because I knew Professor Spemann personally, and surprisingly well owing to a very fortunate circumstance for me. Following my graduate study I enjoyed a post-doctoral fellowship in his laboratory in Freiburg im Breisgau, Germany. This was 1935/36 when he was preparing the English edition of his book on embryonic induction. Shortly after my arrival in Freiburg, Professor Spemann asked me to assist with the revision of a translation of his German text (*Experimentelle Beiträge zu einer Theorie der Entwicklung*). The sentences were long and involved, and I spent many hours re-working them. After the completion of two or three chapters we would have a conference, usually after tea with our wives at his home on the Lorettoberg, a wooded knoll on the outskirts of Freiburg. Without these occasions I would not have known "Den Chef" as a friend; our meetings otherwise would have been infrequent and formal.

In recent years I have been distressed by reading some disparaging comments about Spemann's politics, his competence as a zoologist, and his part in the discovery of the "Organizer." My rejection of these criticisms has been strengthened by correspondence with my good friend Viktor Hamburger, who was one of Spemann's students and later a close colleague. Professor W. M. Copenhaver, a guest in Spemann's laboratory at an earlier time, complimented me last year upon the faithfulness of my portrayal of "Spe" (as he was affectionately known by his assistants and students) in appearance, manner, and speech.

The Spemann Lecture

[Spemann enters in formal academic dress: cutaway coat, striped trousers, high collar, and top hat. He is about sixty-eight; his hair is gray; his moustache is trim. He carries a book, the German edition of his *Embryonic Development and Induction*. He speaks in a soft, gentle voice and with a pronounced German accent.]

Meine Damen und Herren. It is great joy to begin my lecture with ladies and gentlemen instead of by *Heil Hitler*, as we must say in my country these days. I am sorry I do not speak English very well, although I studied your language already beginning in the *Gymnasium* in Stuttgart, my city of birth. "Do you know where Stuttgart lies?" asks an old German verse taught to me by a nursery maid. *"Wisst ihr auch, wo Stuttgart liegt? Stuttgart liegt im Tale, wo's so schöne Mädle gibt, aber so brutale."* In translating *"brutale"*, I suggest that you do not say: beautiful maidens of Stuttgart are brutal. You should say: beautiful maidens of Stuttgart are devastating [fig. 16]. I know from experience. I read English easily. My favorite American authors are Whitman, Emerson and Thoreau. I have travelled a little in England and America. I gave the Croonian Lecture before the Royal Society in London in 1927 and the Silliman Lectures at Yale University in 1933. The Silliman Lectures were the basis of this book on embryonic development. [Spemann picks up the German edition of the Silliman Lectures.] In it I have summarized many of the experiments performed by my

FIG. 16.   *The maidens of Stuttgart are devastating.*

students, associates, and me. Only three studies will be discussed in this lecture.

The first one is on the separation of the first two cells of a salamander embryo. This was done at the University of Würzburg, where I took my doctor's degree under the great Theodor Boveri, who had a profound influence on me. Boveri was not only a fine scientist but also a very kind man. I remember especially his solicitude at the time of my doctor's examination, a very formal occasion in the German universities of that time. After finishing one's thesis a doctoral candidate must appear before his examiners in full dress with tails, striped trousers, white gloves, and–and–*Wie heisst Zylinderhut auf English?* [fig. 17.][Help comes from the audience.] Top hat! *Jawohl.* Top hat. Very colorful word. So. The candidate arrives at ten o'clock in the morning and waits alone until his examiners assemble and finish their conversation. It was then

FIG. 17.   *Wie heisst Zylinderhut?*

that Boveri came to me with words of comfort. After two hours the examination is broken for lunch, which is served to the professors at the expense of the rektor of the university. All doctoral examinations are open to any professor, but most of them come only for the free sandwiches and wine. Meanwhile, the unhappy candidate does not dare to touch the wine, because the rest of the examination is yet to come. If the candidate passes, the next morning he must put on his formal clothes again and call upon each of his examiners at their homes.

At that time there was a controversy in biology between two experimentalists, Wilhelm Roux of Halle and Hans Driesch of Leipzig, regarding the basic character of development. Was the embryo a mosaic of predetermined parts, as claimed by Roux, or was the embryo indeterminate and regulative, the position of Driesch? Roux had worked with the two-celled embryo of a frog. He killed one of the cells by pricking it with a heated needle. The surviving cell formed only a half of an embryo [see figs. 18, 19]. According to Roux this was because the nucleus of that cell had the determinants for only a half of the embryo. Driesch, on the other hand, obtained quite different results with an experiment on two-celled embryos of the sea urchin. By vigorous shaking of a tube of sea water containing embryos in the two-cell stage, he succeeded, in a few instances, in breaking the egg membrane which allowed the two cells to separate from each another. Each developed into a whole embryo [figs. 20, 21]. Who was correct? Roux or Driesch?

I took up the problem in Würzburg after my doctor's degree. I used a still different technique on the two-cell embryo of a salamander which, like the frog embryo, lies inside a membrane and several layers of jelly. I made a fine noose of baby's hair which I stole from the head of my first son. The noose was slipped over the jelly around a two-cell embryo and

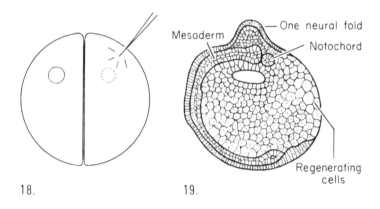

18. 19.

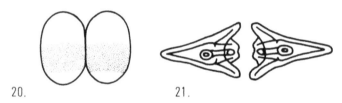

20. 21.

FIG. 18. *Roux killed one cell in a two-cell embryo of a frog.* FIG. 19. *Result: a hemi-embryo.* FIG. 20. *Driesch separated the first two cells of a sea urchin.* FIG. 21. *Result: twin larvae.*

then slowly tightened, constricting the embryo until the two cells were separated [fig. 22]. Each cell developed into a whole embryo. Thus, I had formed a pair of twins from one egg [fig. 23], in agreement with Driesch. Accordingly, both echinoderms and chordates are indeterminate and not mosaic in their development. Later, in 1935, when I received the Nobel Prize for another discovery, about which I shall speak in a moment, there was considerable embarrassing publicity in the newspapers, and one paper, the *Berliner Illustrirte Zeitung,* called its article *"Zwillinge nach Wunsch"*–"Twins as You Wish."

[Spemann holds up a copy of the *Berliner Illustrirte* showing photographs of the noose experiment.]

But it turns out that nature is not always so simple as one thinks or would like it to be, as I discovered in my noose experiments. Sometimes, I did not get twins. Why? I soon discovered the secret. There is a region on a fertilized amphibian egg—frog or salamander—called the gray crescent. This had been known for some time. It is a lightly pigmented, crescentic area which appears near the equator of the egg shortly after fertilization. Usually the first cleavage cuts the gray crescent so that each of the first two cells gets a part of it [fig. 22]. But sometimes the first cell division divides the fertilized egg so that one cell receives all of the gray crescent and the other has none [fig. 24]. I discovered that if the noose constricted an embryo in the first instance (sagittal cleavage) I obtained twins, but in the second type (frontal cleavage) there developed one normal embryo and a *Bauchstück—Wie heisst Bauchstück?* [Help comes again from the audience.] Yes, belly-piece [fig. 25]. A belly-piece is made of skin and gut but it has no notochord, no neural tube, no eyes, no ears, *und so weiter*. Thus you see that the gray crescent is vital for the normal development of an embryo. I have often thought that I should write a love song entitled, "Without my gray crescent to embrace me I am just a belly-piece." [Spemann makes a large crescent with outstretched arms which close in a hug.]

Also. The amphibian embryo is not so completely regulative and indeterminate as suspected. It has a touch of mosaicism. The grey crescent is a special region. It must contain something essential for the development of notochord, nervous system and sense organs. I did not fully understand these results until several years later when I discovered the organizer, which I shall describe later.

Now, not only can a whole embryo develop from half of an egg, but a whole embryo can develop from two eggs fused together. Take two eggs after first cell division, remove the jelly and egg membrane from both, and place one on top of the other. They will fuse together, continue to divide, form a ball of cells which develops into a normal embryo of giant size. One can even fuse the embryos of two different species of sala-manders and obtain a chimera, that is, an embryo which is normal in every respect but constructed from cells belonging to two different species, as shown by two of my colleagues [figs. 26, 27]. The power of nature to heal and to regulate and to overcome adversity is fantastic.

FIG. 22.   *Noose experiment applied to a two-cell embryo with sagittal cleavage.* FIG. 23. *Result: twins.* FIG. 24. *Noose experiment applied to a two-cell embryo with frontal cleavage.* FIG. 25. *Result: one whole embryo and a belly-piece.*

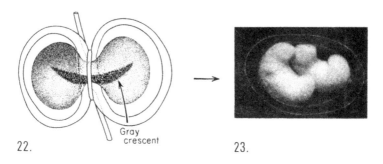

22.

23.

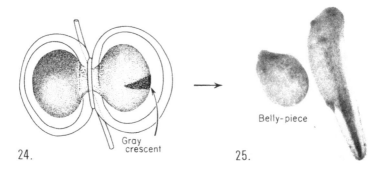

24.

25.

I now discuss a second work–on the development of the vertebrate eye. These experiments were also conducted at Würzburg when I was a *Privatdozent*, which position corresponds to your lecturer. The *Privatdozent* in the old German system, however, did not receive a salary. He collected fees from the students who attended his lectures. The professor, only one in each field in a university, always lectured in the large general courses, so he collected the largest number of fees, in addition to his salary. A *Privatdozent* lectured to small classes and consequently received few fees. Indeed, he must have another source of revenue or starve.

For transplantation of small parts of amphibian embryos special techniques were needed. I used very simple instruments [fig. 28]: glass-needle knives to cut the soft embryos, a loop of baby's hair mounted in a glass capillary tube to manipulate an embryo, balled glass rods to make depressions in wax to cradle an embryo, and tiny glass bridges to hold grafts in position after they have been transplanted. I never ceased to marvel at the beauty of an embryo, how tender yet how swift to heal and repair any damage inflicted upon it. For hours I could sit and watch an egg cleave, or the cells move through the blastopore, or the neural folds rise and fuse to form a neural tube, or the

FIG. 26. *Fuse two early embryos from two species; cells (1) and (2) from* Triturus alpestris *and cells (3) and (4) from* Triturus taeniatus. FIG. 27. *Result: a whole embryo, in neurula stage, composed of cells from the two species.*

26.   *(2)*   27.

tailbud embryo move in circles within its membrane by the beat of thousands of cilia. [Spemann becomes mystical.] Some scientists see the beauty and order in the universe and the handiwork of the Creator in the structure of the atom, or in the sweep of the Milky Way and the swing of the Pleiades, or in the homeostasis of blood sugar, or in the adaptations of birds for flight, or in the color and patterns of insects; but I found them in the embryo.

And what a glorious society we would have if men and women would regulate their affairs as do the millions of cells in the developing embryo. We should study nature for moral lessons. This injunction has been better stated many times before. Bryant wrote in "Thanotopsis": "Go forth under the open skies and listen to the teachings of nature. She speaks a various language." *Auf Deutsch, en français,* in English. And Shakespeare said: there are "tongues in trees, books in running brooks, and sermons in stones." And in the Bible we read: "Go

FIG. 28. *Microsurgical instruments used in amphibian experimental embryology.*

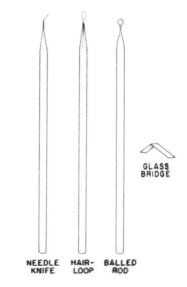

to the ant, thou sluggard. Consider her ways and be wise."
[Spemann realizes that he has drifted away from his subject.]
But now we go to the embryo.

I learned to remove and to transplant the embryonic
eye—called the optic vesicle—which forms as an outpocketing,
like a balloon, from each side of the embryonic brain. When
this optic vesicle grows to contact the skin ectoderm, the outer
covering of an embryo, it invaginates to form a cup, called the
optic cup. [Spemann makes two neat sketches on the chalkboard, one showing an optic vesicle and the other an optic cup,
figs. 29, 30.] The inner thick layer of the cup becomes the
retina. The lens of the eye, however, arises not from the out-
growth of the brain but from the skin ectoderm. [Spemann
adds the lens to the second figure.]

We are now ready for some experiments. Remove an
optic vesicle at an early stage in its development in a frog
embryo by cutting the optic stalk—so. [Spemann draws a line
AB across the optic stalk in the first diagram.] Result: not
only no eye, but also no lens [fig. 31]. Without the stimulus
provided by the optic vesicle, the skin ectoderm will not form a
lens. To be sure of my conclusion, I conducted another experiment. I transplanted a young optic vesicle beneath some *belly*
ectoderm. A beautiful lens was there formed by the ectoderm
[fig. 32]. Normally, of course, lenses do not grow on bellies.
[Spemann chuckles.] But given the stimulus, whatever it is,
from an optic vesicle, skin ectoderm of a frog embryo responds
with the proliferation of a mound of cells that transforms into
a crystalline lens. Some qualifications about age and species of
frog embryo used might be added, but they would not alter the
basic principle I have just illustrated. I called this principle
induction.

As you may be acquainted with my work on the dorsal lip
organizer, I shall be brief about my third topic. By dorsal lip I

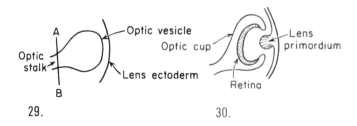

29.

30.

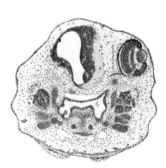

31.

32.

FIG. 29.   *Optic vesicle stage in development of vertebrate eye.* FIG. 30. *Optic cup stage in development of vertebrate eye.* FIG. 31. *Result of removing left optic vesicle: no eye and also no lens on left side.* FIG. 32. *Result of transplanting an optic vesicle under belly ectoderm: transplant forms an eye and induces a lens in the belly ectoderm.*

refer, of course, to the upper lip of the blastopore, the opening in the early embryo through which cells move to form future internal organs. Those cells rolling in dorsally give rise to skeleton and muscle. More importantly, however, they determine the destiny of the overlying external cells, inducing them to become brain, spinal cord, eyes, and ears. So decisive is its inductive action that I termed this dynamic center of the embryo the primary organizer. It was this work which won the Nobel Prize in Medicine and Physiology in 1935, although the crucial experiment was done much earlier, in 1925, in collaboration with my graduate student and research assistant, Hilde Mangold. In that experiment we transplanted the dorsal lip from an early embryo (donor) of one species of salamander into the side of another embryo (host) of the same age but of a different species [fig. 33]. By cross-transplanting between two species, which differ in size of cells and amount of natural pigment, it was possible to distinguish between donor and host cells at a later stage of development. The transplant organized (induced) a new embryonic center on the side of the host. The former I called a secondary embryo, the latter the primary embryo [fig. 34]. In rare instances, a secondary embryo may develop into an almost perfect individual attached to the side of the host [fig. 35]. Now it became clear why the gray crescent region of the fertilized egg is so important for development. Without it, you remember, only a belly-piece forms. The reason was now clear: the gray crescent marks the future dorsal lip area; hence, it is the forerunner of the organizer.

In the last chapter of my book I discuss some general topics including my ideas on how the good scientist should work. I have found an analogy to an archeologist very helpful. Let me read a paragraph to you as it will appear in the English edition of my book. [He removes a card from the German edition and puts on a pince-nez.]

I should like to work like the archeologist [fig. 36] who pieces together the fragments of a lovely thing which are alone left to him. As he proceeds, fragment by fragment, he is guided by the conviction that these fragments are parts of a whole which, however, he does not yet know. He must be enough of an artist to recreate, as it were, the work of the master, but he dare not build according to his own ideas. Above all, he must keep holy the broken edges of the fragments; in that way only may he hope to fit new fragments into their proper place and thus ultimately achieve a true restoration of the master's creation. There may be other ways of proceeding, but this is the one I have chosen for myself.

FIG. 33.   *Transplant dorsal lip organizer from one gastrula (donor) to ventral side of another gastrula (host).*   FIG. 34.   *Result: transplant organizes a secondary embryo.*   FIG. 35.   *Sometimes a secondary embryo is as well-developed as the primary embryo.*

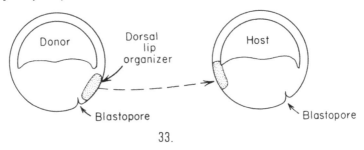

33.

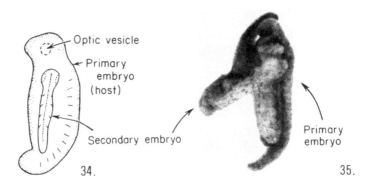

34.   35.

FIG. 36. *I should like to work as the archeologist.*

Perhaps I can make my point better with a demonstration.
I have here two inexpensive vases which I bought in one of your
5- and 10-cent stores. Incidentally, you should call them dollar
stores. This one is unbroken. But this one has been smashed.
Perhaps kitty knocked it off my desk. [Spemann scatters the
pieces on the podium.] Suppose—just suppose—that these are
very valuable vases or that they hold dear memories for me. I

should like to restore the broken one. So I take the pieces to an artisan. And I tell him: My good fellow, please put my vase together. Restore the thing of beauty. To be faithful to my analogy with the archeologist, I must not give the artisan the unbroken vase. [Spemann places the whole vase to one side.] He has only a handful of shattered pieces. He does not know what the whole vase is like. And so he begins by trial and error

FIG. 37.   *One must keep holy the edges of the fragments.*

to fit the pieces together. [Spemann picks up two pieces of the broken vase, fig. 37.] Now here is the critical point. He must not work according to his own ideas. That is, he must not force pieces together. He must keep holy—*heilig*—the edges of the fragments. If the pieces match, they will fit together without alteration and without pressure.

And so the good scientist works, with fragments of truth, piecing them together carefully and critically, to bring forth the whole truth, the thing of beauty. "Beauty is truth, truth beauty," said Keats. As he works he must have no convictions of his own—hypotheses yes, but no conclusions. No preconceived ideas, no prejudices, and no allegiances which will impair the objectivity of his research. This is what I mean by inner academic freedom. When I sit and reflect upon beauty and harmony in an embryo I am Hans Spemann the philosopher, I am not Hans Spemann the scientist. In the white heat of research or in the critical evaluation of the work of another scientist matters of religion, politics, social position and personal gain are to be rigorously excluded.

Now I go one step further. Not only must the scientist have academic freedom within himself, but he must be free from all outside pressures as, for example, those from government. In my country at the present time *Akademische Freiheit* is being destroyed. It is my sincere wish that you and your professors will never be subjected to this kind of tyranny. And so, ladies and gentlemen, it has been a privilege to address you. *Danke schön und aufwiedersehn.* [Spemann picks up his book and top hat, nods, and smiles, and leaves the auditorium.]

G. M.

Gregor Mendel

1822–1884

Mᴀ Y INTERPRETATION of Pater Mendel as a jolly priest is based largely upon his taste for secular pleasures, two of which were food and cigars. There were others. I am indebted to my colleague, Curt Stern, for advice and for information beyond that available in the literature. His book with Eva Sherwood, *The Origin of Genetics—A Mendel Source Book*, contains not only a fresh and more accurate translation of Mendel's papers and letters to Carl Nägeli but also other important post-Mendelian documents. To set the stage for the following act, I quote the first papagraph from Stern's foreword to this book.

Gregor Mendel's short treatise "Experiments on Plant Hybrids" is one of the triumphs of the human mind. It does not simply announce the discovery of important facts by new methods of observation and experiment. Rather, in an act of highest creativity, it presents these facts in a conceptual scheme which gives them general meaning. Mendel's paper is not solely a historical document. It remains alive as a supreme example of scientific experimentation and profound penetration of data. It can give pleasure and

FIG. 38.   Grüss Gott.

provide insight to each new reader–and strengthen the exhilaration of being in the company of a great mind at every subsequent study.

Some of my friends in botany who have heard this lecture or the film version of it claim that I give Mendel greater insight into genetics than he deserves. Perhaps–especially when "Mendel" says that someday a powerful microscope might be invented which will permit us to see the *Elementen* and hopefully a chemist will someday inform us of the chemical compostition of the units of heredity. And, of course, "Mendel" uses anachronisms such as a Punnett square to show the genotypes resulting from a dihybrid cross. But I remind the reader that the purpose of these lectures is instruction of students in a course in elementary biology.

I acknowledge the assistance of Dr. John Strother of the University of California Botany Department and Herbarium on botanical matters, but I accept full responsibility for the fraud in "Mendel's" red and white sweet peas. Although "Mendel" places the flowers in jars with water, they are plastic. Moreover, when I conceived the idea of using plastic flowers (always fresh and available), I found that the markets in San Francisco had only pink ones. But red and white paint applied by an assistant, Merianne O'Grady, solved that problem. And I am grateful to Father Francisco Vicente, O.P., of St. Mary Magdalen Church of Berkeley, for the loan of the crucifix and a capuche which was copied by another assistant, Carol Mortensen.

The Mendel Lecture

[Mendel enters the auditorium dressed as an abbot in a long black clerical gown and a capuche over his head (fig. 38). He wears nineteenth century spectacles with oval lenses, narrow frame, and slender temples; a crucifix hangs from his neck. He is about fifty-five years of age and very stout. He carries a small book, a volume containing his classic paper on inheritance in peas published in *Verhandlungen des naturforschenden Vereines in Brünn* (1866). He is smoking a cigar. He walks briskly to the podium, puts the book and cigar aside, tosses back the capuche, and smiles warmly to his audience. He speaks with a heavy Germanic accent.]

Grüss Gott. My lecture today is on inheritance in hybrids of the common edible pea, genus *Pisum.* My interest in inheritance in plants and animals began already as a boy on my father's small farm in Heinzendorf in Moravia, a province of Austria. My father had horses and cows, chickens and bees, peas and beans, and always flowers. Being a curious and uninhibited boy I observed breeding of animals and seed formation in plants, and I often wondered why the offspring resembled their parents but were never exactly like their parents. We had a good teacher in Heinzendorf school, which had been started only twenty years earlier. Previously the children of Heinzendorf never learned to read or write. The people were poor peasants and could not afford to send their children away to school. From my teacher I learned much, including growing fruits and keeping bees.

When I completed the school my teacher told my parents that I should go to high school in a town about twenty kilometers distant. With great sacrifice to them I went, but it was hard to keep body and soul together. Part of the time I was on half rations. Eventually, it appeared that I must abandon my studies and begin to earn my livelihood. But one of my high-school teachers recommended me to the Augustinian Monastery of St. Thomas at Brünn. I was accepted and became a novice, and later I took my vows in accordance with the rule of St. Augustine. After a few years my order sent me to the University of Wien [Vienna] with hopes that I would pass the examinations and become accredited as a teacher. Well, I did not pass my examinations, and so I returned to Brünn and became an unaccredited teacher—a good one, I am happy to say. I always kept animals, especially bees and mice and also bright flowers. But I did not like snakes. Yet my boys brought them to me because they knew that I did not like snakes.

My interest in plants and animals, which began on my father's farm as I have explained, continued during my school years and later study at the University of Wien, and became my primary preoccupation at the monastery in Altbrünn when I was not engaged in teaching or religious duties. My animal breeding in the monastery, however, was regarded as immoral by my superiors because it appeared that I was playing with sex. I had to be particularly careful not to incur further disfavor of the bishop, a conservative cleric and a very corpulent man. I had made the mistake of remarking to some friends once that the bishop carried with him more fat than understanding. Someone betrayed me and reported me. Thereafter, the bishop and I were not good friends. Incidentally, little did I know at that time that I too would acquire stout proportions. [Mendel pats his stomach.] So I turned from animal breeding to plant breeding. You see, the bishop did not understand that plants also have sex.

I had long grown ornamentals for their lovely flowers, and I had learned to cross-pollinate them to obtain interesting hybrids. But these plants were not suitable for a study of inheritance. After trying several kinds of plants I found the common edible pea most satisfactory, for several reasons. First,

FIG. 39. *Reproductive parts of the garden pea* Pisum sativum; *other petals (two wings and keel) have been removed.*

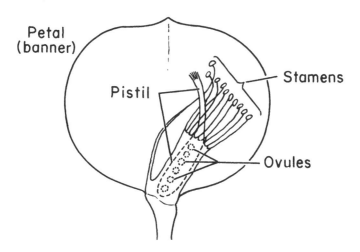

FIG. 40. *And peas are nutritious.*

seeds of many varieties could be easily obtained from seedmen in the town. Second, the plant grew well in the monastery garden, that is, in the small plot assigned to me. Third, its flower was ideally constructed for experiments on hybridization. Normally the plant reproduces by self-pollination. But this can be prevented by removing the stamens before the pollen is ripe. The flower is like this. [Mendel draws a sketch on the board, fig. 39] Simply spread apart the petals, and then you see the sex organs inside. Pull out the stamens, the male structures, with a pair of tweezers, leaving the pistil, the female organ. Cross-fertilization can then be accomplished by dusting pollen from a ripe plant on the pistil by means of a small brush. The petals enclose the pistil so well that airborne pollen cannot enter, and most insects likewise. Thus, the chances of contamination with foreign pollen are slight, but I never relied on it. There is the pesky pea weevil which can invade the bedroom chamber. So I always tied bags of paper over the flowers in my experiments. And four, there are other advantages to peas. They are nutritious. I usually carry a pocketful of them to satisfy my yearning. [Mendel produces a handful of peas from his gown and pops some into his mouth, fig. 40.] And when my boys pay no attention to my lectures or fall asleep, I simply ask the Almighty to send down a shower of peas. [Mendel peppers his audience with peas.] That wakes them up! *Gelt?*

After a few years of preliminary studies to test the purity of my seeds and to select features for study I began my experiments in earnest in 1857 and finished them eight years later, after studying over 10,000 plants. I presented my results in 1865 to the Brünn Society for the Study of Natural Science, and they appeared in the *Proceedings* of our society in the following year (1866).

I found that some features had clear contrasting traits such as tall and short vine, six to seven feet versus three quarters to one and a half feet.[1] Or another example: the color of flower, either red or white. In all I selected for study seven features with clear and distinct traits. I avoided those features which were indistinct. The seven I chose were color of seed, shape of seed, color of pod, form of pod, color of flowers, position of flowers on the stem, and length of stem. For each of these features one of the traits was dominating and the other was recessive. Let me illustrate what I mean by dominating and recessive.

[1] According to Stern and Sherwood (1966), Mendel probably used a Viennese foot which is equal to 0.316 meter, or 12.44 inches.

I have here some red and white sweet-pea blossoms. [Mendel picks up a bowl of plastic sweet peas of the two colors.] Let us assume that they are flowers of the garden pea. I cross a red-flowering plant with a white-flowering one. [Mendel playfully rubs together red and white plastic blossoms and looks up at his audience with a grin, fig. 41.] It makes no difference in the results whether I dust the pistil of the red plant with pollen from the white one, or vice versa. All of the next generation have red flowers, one hundred percent—not a single white one. [He places a red flower in a glass jar behind those holding the red and white parents.] I use just one to symbolize the many. Now, is the white trait lost? No! When I cross two of the first-generation plants, lo and behold white flowers will appear in all their purity. [He places a white plastic sweet pea at the third level on the lecture table.] The whiteness was only supressed by the redness in the first generation offspring, and that is why I called white recessive and red dominating.

Of course, there are also red-flowered plants in the second-generation offspring, and when I counted them there were about three of the reds for every one of the whites. [He places three red flowers in a container adjacent to the one with a white flower.] You see that the ratio of the reds to the whites is about 3 to 1. Now, I say "about"—it was never exactly 3 to 1. As examples, I quote from my paper. [Mendel picks up the volume containing his first publication on inheritance.] In one experiment I had 929 plants: 705 had red flowers, 224 white ones. The calculated ratio was 3.15 to 1. In another instance, the ratio was 2.95 to 1. I think that if I had produced more plants my ratios would probably be closer to the theoretical 3 to 1.

I now proceed to analyze the second-generation plants. We discover, not unexpectedly, that the white plants when crossed with white give white. Actually, it is easier to allow a white pea to self-pollinate. Thus pollen from a white plant fertilizes ovules from a white plant. Obviously, the whites are a pure line. Consider now the red ones: one third of them turn out to be pure red-breeding. When crossed with each other or when you allow self-pollination they produce only red-flowered plants, just like grandfather. The other two-thirds of the second-generation reds are like father and mother, that is hybrid red, because they give a 3 to 1 ratio of red to white upon inbreeding. So we actually have in the second-generation plants three kinds: pure red, hybrid red, and white. And the ratio between the kinds is 1 to 2 to 1. *Gelt?*

FIG. 41.   *I cross red and white flowered peas.*

I began to think of discrete hereditary units–I called them *Elementen*–which pass from generation to generation by way of the pollen grains and the ovules. And there must be an element for redness–a dominating element–and an element for whiteness–a recessive element. When they are together in a hybrid, the flowers are red because the element for redness dominates the one for whiteness. But the element for whiteness is there, all the time. It will not be lost or contaminated. I then devised a simple system for recording the elements, to save much writing. I said, let capital *A* stand for the dominating element and small letter *a* for the recessive one. The hybrid red, since it has both *Elementen*, should be expressed as *Aa*. Now when such hybrids produce pollen grains and ovules the elements separate from one another so that in the stamens two kinds of pollen are formed; half of them will possess large *A* and half of them little *a*. Likewise, in the pistil two kinds of ovules will develop: those with *A* and those with *a*. This separation of the *Elementen* I called the law of segregation.

You must understand that I was not the first person to study inheritance. Man has been breeding plants and animals since ancient times, and many scientists have tried to discover the principles of heredity. But why did I succeed and the others fail? One reason is just this: I selected a few well-defined, contrasting traits to study in an organism which was suitable for experimentation. My predecessors, however, tried to study the whole organism at once. But inheritance is too complex. One must simplify the investigation; study one pair of contrasting traits at a time. And then you can study two traits simultaneously, as we shall do in a moment. Only in that way can one discover the secrets of nature. Then second, I used quantitative methods. I counted peas, and peas, and peas! I kept very careful records, and I calculated ratios. During the long winter months I worked on the seeds, counting them, placing them in properly labeled envelopes, posting my ac-

FIG. 42.   *So, there are four kinds of pollen.* Gelt?

counts, doing the calculations, and planning the experiments for the next spring. Hard, tedious labor, but that was how I succeeded. I shall not say that my predecessors were lazy. Let me only claim that I was willing to work hard and with the persistence and stubbornness of my peasant ancestry.

So, having studied all seven of the contrasting traits and learning that they obeyed the law of segregation, I went on to study two and three traits simultaneously. Let me illustrate. Suppose we cross a pure-breeding red-flowered tall pea plant with a white-flowered short one. [Mendel writes on the board: *AABB* × *aabb*.] Red is dominating over white, as we have seen, and tall is dominating over short. Note that I use large *B* and small *b* to represent the *Elementen* for tall and short, respectively. All of the first-generation offspring will be hybrid red and hybrid tall or *AaBb*. Gelt? This follows from the law of segregation. Now let us cross two such hybrids or simply allow self-pollination. I then discovered my second principle, the law of independent assortment, which means that the *Elementen* for one trait (color of flower) segregate independently of the *Elementen* for another trait (length of vine). [Mendel demonstrates the four types of pollen and ovules which would be formed by the use of arrows on *AaBb*, as shown in fig. 42.] Next, let us analyze the second-generation offspring. [Mendel uses a Punnett square, fig. 43, to demonstrate the ratios of the four phenotypes and the nine genotypes. He then explains the arithmetic progression in numbers of gametes (pollen and ovules), phenotypes (traits), and genotypes (kinds), and combinations of gametes in crosses of hybrids, using fig. 44.]

Also! What are these discrete elements about which I have been speaking? My microscope [fig. 45] does not greatly magnify. I can see the tiny pollen grains easily but not the *Elementen* inside them. I hope that someday a powerful microscope will be invented so that we can see the elements of

OFFSPRING RESULTING FROM AaBb x AaBb

Ovules	AB	Ab	aB	ab
Pollen AB	AABB	AABb	AaBB	AaBb
Ab	AABb	AAbb	AaBb	Aabb
aB	AaBB	AaBb	aaBB	aaBb
ab	AaBb	Aabb	aaBb	aabb

FIG. 43. *Second generation offspring from a cross between a pure breeding red tall pea plant and a white short one.*

HYBRID CROSSES

NUMBERS OF :					
Features	1	2	3	4	n
Traits and Types of Pollen & Ovules	2	4	8	16	2^n
Kinds	3	9	27	81	3^n
Combinations of Pollen & Ovules	4	16	64	256	4^n

Equivalence of terms used by Mendel with modern ones:

Feature . . . Character	Dominating . . . . . Dominant
Trait. . . . . Phenotype	Recessive. . . . . Recessive
Kind. . . . . Genotype	Pure dominating . . Homozygous dominant
Element . . Gene (in general)	Pure recessive. . . . . Homozygous recessive
Allele (member of a pair)	Hybrid . . . . Heterozygous

FIG. 44.   *Arithmetic progressions in hybrid crosses.*

inheritance, and perhaps clever chemists will be able to tell us their chemical composition. Much more work needs to be done on inheritance.

In a long conversation with some visitors recently in the monastery garden, my mind was stirred anew on the subject of evolution. One member of the group, a botanist, asked: "What is the relationship, if any, between your work and the origin of new species?"

"I am convinced," I replied, "that my studies on hybridization are of significance for the evolutionary history of organic form."

They pressed me further: "Was I acquainted with the writings of Charles Darwin?"

"Yes, indeed," I said. "I have read almost everything that he has written. The library in our monastery contains most of Darwin's books and also *Zoonomia* by his grandfather, Erasmus."

I have long thought that there was something lacking in Darwin's theory of natural selection. And I have also questioned the views of Lamarck. I once made an effort to test the influence of environment on plants. I transplanted certain plants from their natural habitat to the monastery garden in an attempt to produce permanent variations. Although cultivated side by side with the form typical of the garden, there was no approximation between the two. No change occurred in the transplanted forms as a result of the change in environment, even after several years. Nature does not modify species in that

FIG. 45. *Mendel's microscope.*

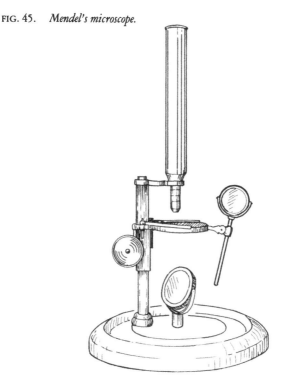

way, so some other forces must be at work. My visitors then asked me about new experiments which I was conducting.

I sadly looked about the garden. Not one of my plants could be seen. All were gone. I have become so burdened with the administration of the monastery that I no longer have time for experiments. And I have become discouraged. Few appear to know about my work—not even Charles Darwin. And those with whom I am in correspondence? They do not seem to understand my experiments—not even some leading German biologists. But I take comfort, however, in the knowledge that science is always moving forward, even though slowly at times. Sooner or later—sooner or later—my work will be tested, and I shall be either right or wrong. In the meantime I must have patience. I believe that my day will come. That is what I tell my boys. I say to them: "Work hard, work with joy, and work with patience." And I often commend to them a verse in the Book of Ecclesiastes:

"The patient in spirit is better than the proud in spirit. Therefore, be not hasty, and be not angry: for haste and anger rest only in the bosom of fools." May God bless you.

[Mendel picks up his book, returns his cigar to his mouth, smiles and nods, and quickly leaves the auditorium.]

L. P.

Louis Pasteur

1822–1895

IN DEVELOPING this series of lectures for classes in Zoology 10, I followed the advice of Louis Pasteur: study great men. This injunction he delivered to undergraduates at the University of Edinburgh in 1884 when he attended the tercentenary celebration of that renowned Scottish university as a delegate of the French Academy of Sciences. Little did Pasteur know, probably, that he would become one of the distinguished biologists of history and an example to many generations of students.

As a teenager I acquired an abiding esteem and affection for Pasteur, initiated by a reading of Paul de Kruif's *Microbe Hunters*, a book for which I now have an aversion. Then I read *The Life of Pasteur*, by René Vallery-Radot, Pasteur's son-in-law (he married Marie-Louise Pasteur). A third contribution to my early appreciation of the great French microbiologist was made by Paul Muni's motion-picture characterization.

I added "Pasteur" to my repertoire of impersonations for my course four years after the other lectures were prepared. The delay was owing to a sense of inadequacy to portray the man whom the *Spectator* eulogized, upon his death, as the most nearly perfect person "who has ever entered the Kingdom of Science." Secondly, I wished to deliver the lecture with a French accent, but foreign languages have always been difficult for my sluggish tongue. I am pleased to acknowledge the gracious assistance on the latter problem of my good friend Dr. Armelle Kernéis of the Faculté des Sciences, Université de Paris. I sent an early draft of this lecture to her to read aloud into a tape recorder. The director of the movie version of this lecture, John Quick of the Lawrence Hall of Science, was also helpful. Incidentally, Christine Sykes Happ of Fort Collins, Colorado, is similarly helping me to speak the Queen's English in next year's performances of "Darwin" and "Harvey."

In addition to biographies, principally Vallery-Radot's, and Pasteur's own writings, I have used J. R. Porter's article in *Science* published in celebration of the sesquicentennial of the birth of Pasteur. The cover of this issue of the journal (December 22, 1972) reproduces in color one of Pasteur's paintings, that of Father Gaidot, a cooper who was a neighbor of the Pasteur family. According to Professor Porter, "Art critics even today say Pasteur could have made a great reputation for himself in the arts." He received a certificate in arts in 1840 in Besançon and considered painting as a career. Although captured by science, Pasteur, unlike Darwin, retained a lifelong interest in the fine arts. From 1863 to 1867 Pasteur was a professor of geology, physics, and chemistry in relation to the arts at the Ecole des Beaux Arts in Paris.

Two of the exhortations to students included in this lecture were taken from Pasteur's speeches, one from an address as a young professor and dean of the new Faculté des Sciences at Lille, the other from his valedictory to the impressive audience assembled at the Sorbonne on December 27, 1892, to honor him on his seventieth birthday. But I must stop talking about Pasteur and begin speaking as *Monsieur le Docteur et Professeur Louis Pasteur*.

The Pasteur Lecture

[Pasteur enters the lecture hall on the arm of an assistant who wears a white laboratory coat. Pasteur is dressed in black trousers, dark blue great coat, high white collar, and narrow black tie. A pair of spectacles dangles from the right lapel of his coat; the left lapel bears a red ribbon – the decoration of the *Grand Croix de la Légion d'honneur*. He carries a cane and limps because of a left hemiplegia, the result of a stroke at the age of forty-six. He is a striking figure. His hair is mostly black, his temples and short beard gray, and mustache and goatee white. Pasteur walks slowly to the podium with the assistant, places his cane on the lecture table, thanks the assistant, "*Merci beaucoup*, Mousieur Baylis," and addresses the class.]

Mesdames, Messieurs: It is an honor to speak to this magnificent audience of students eagerly preparing for the great adventure of life. *La vie! Oui, la vie*. Life! That is the theme of my lecture today. Life in all creatures, large and small, saints and microbes. What is life? Where did life begin? And how did life begin? My scientific work, however, did not commence with these questions. I was first a chemist and crystallographer. I studied the symmetry of molecules, such as the right-handed and left-handed molecules of tartaric acid, which are mirror images of each other, as I found by examining crystals of tartaric acid under a microscope. [Pasteur points with his cane to the dextro and laevo crystals sketched on the chalkboard, fig.

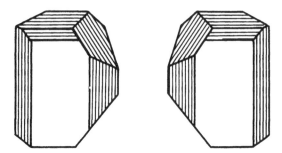

FIG. 46.   *Right-handed and left-handed crystals of tartaric acid.*

46.] But the molecules of tartaric acid, they are not living! Then how did I, a chemist, become a biologist? One day, I discovered that a certain bacterium ferments only the right-handed molecules of tartaric acid, leaving the left-handed ones to accumulate. Only right-handed molecules were food. What a discriminating microbe! *N'est-ce pas?* Later I studied the production of other acids: acetic, lactic, butyric, et cetera. I soon learned that different microorganisms give different fermentation products. Yeasts, for example, produce alcohol, whereas certain kinds of bacteria form lactic acid, as in the souring of milk, and still others make butyric acid, as in rancid butter, and so forth. You see that I was slowly becoming a biologist without at first knowing it, for fermentation is associated with life. Indeed, living organisms are essential for fermentation. In recognition of my studies on the symmetry of molecules I was awarded the red ribbon of the *Légion d'honneur* [Pasteur touches his left lapel reverently]—the same decoration which my dear father won for gallantry under Napoleon I in the service of France.

These early investigations on fermentation were followed by studies on the diseases of wine and beer. Because of these diseases—bitter wines and sour beer—two of France's important industries were facing disaster. Already England was refusing shipment of our renowned Chateau Lafite, Romanée-Conti, Chateau Figeac, and other wines. And the sailors in the French navy were complaining of the vinegar they were forced to drink. [Pasteur chuckles.] So when French industry and the French navy and even the emperor requested my help, I gladly responded. It has always been my firm conviction that the first allegiance of a scientist is to truth, but after that I place my beloved country.

One of the first things that I did was to compare a drop of good wine with a drop of bad wine under a microscope. [Pasteur places samples from each of two wine bottles on a microscope slide and looks at them under a nineteenth-century brass microscope.] The good wine was clear, but the bad wine—it was cloudy, and I saw many particles [fig. 47]. Then I placed a small drop of the bad wine into a bottle of good wine. Incidentally, I always tasted the good wine first to be sure that it was good. [Pasteur pours a sample of Chateau Lafite, tests its fragrance, and then drinks a little, fig. 48.] It is good. [He adds a drop of bad wine to the bottle of good wine.] After a few days the good wine became cloudy and bitter. Bottle after bottle of fine French wine was spoiled by the inoculations with bad wine. Here then was a disease of wines caused by infection. Since the particles multiplied and since reproduction is a characteristic of living organisms, I concluded that the particles were living. They were bacteria causing the wrong kind of fermentation!

Perhaps, I thought, one might be able to kill the undesirable microbes, and thus cure the disease. So I heated the bad wines to 50 or 60 degrees Celsius. Then I repeated the experiment, adding a drop of the heated bad wine to bottles of good wine. I waited. To my great joy and the honor of France, the good wine did not become cloudy and bitter. The French navy repeated my experiment on a grand scale. Five hundred liters of wine destined for the fleet and the colonies were divided: one half was heated, the other half unheated. Both were placed aboard the outbound ship *Jean Bart* which returned to Brest ten months later. The heated wine was limpid and mellow; the unheated wine was acid and barely fit to drink because of the growth of microorganisms.

My heating procedures were also adopted in the beer, vinegar, and milk industries. The Austrians have called the process *pasteurization*. Then I was invited to London to visit the famous British breweries. I called for a microscope and a sample of the yeast they used. *Parbleu!* I observed the fila-

FIG. 47.   *The bad wine was cloudy, and I saw many particles.*

FIG. 48.   *I always tasted the good wine first, to be sure that it was good.*

ments of disease among the globules of yeast. The English, who pride themselves on their fine beers and ales, did not at first believe me. What does a Frenchman know about beer, they thought. But later they admitted that they were having problems with spoiled beer. They agreed to buy a microscope and to be alert to the filamentous growths in their barm and to use controlled heat to destroy them. I returned to Paris happy that I had made new friends for France.

But these undesirable microbes! Where and how did they arise? By spontaneous generation, as some believe? When I began to ask these questions of myself and of my students and colleagues, my close friends said: "Oh no, do not waste your time on such worthless philosophical problems. Many a scientist has floundered and perished in the quagmire of spontaneous generation. Continue your work for French industry. Do you not love your country?" I replied: "It is not a question of love of France. I have served her well, and I shall continue to serve her, but the origin of life is a profound problem." With few exceptions past discourses on spontaneous generation have been metaphysical exercises conducted with great passion, but without adding to our scientific knowledge.

I could not set aside my burning desire to bring a little stone, God willing, to the frail edifice of our knowledge of the deep mysteries of life and death, where all our intellects have so lamentably failed. In defense of nonapplied science I have repeatedly told my students—and I tell it again today to this impressive audience of students—that without theory, practice is but routine. Theory only is able to cause the spirit of invention to arise and develop. It is important that students, above all, should not share the opinion of those who disdain everything in science that has no immediate application. In science, chance favors only the mind which is prepared. I repeat: in science, chance favors only the mind which is prepared.

I first confirmed the experiments of the Italian abbé, Lazzaro Spallanzani, known also for his studies on gastric digestion. I made a nutritious broth, put it in a flask such as this [Pasteur holds up a large flask containing a brown solution], heated it to violent boiling, and then while it was still steaming sealed the neck of the flask in a flame. [Pasteur points to the hermetic seal of the flask.] My results agreed with those of Spallanzani: the broth remained pure. But if the neck be broken to admit air, the broth soon became putrid. The critics of Spallanzani and the critics of Louis Pasteur said that the heating made the air in the flask unfit for spontaneous generation. Only when fresh air is admitted, can life begin anew. I argued in vain—even before our Académie des Sciences—that the putrefaction was caused by admission of bacteria. More convincing experiments were needed.

I opened flasks of sterilized broth in the cellar of the Paris observatory where the air was still. Only one flask out of ten became putrid, whereas eleven flasks out of eleven opened in the courtyard quickly acquired a rich growth of bacteria. I journeyed to Mt. Montanvert in the Alps where I opened twenty flasks of sterilized broth. Only one became putrid. I concluded that the air in the cellar and the air above the *mer de glace* was freer of bacteria than the air in the city streets. But my adversaries performed similar experiments with different results. I think that they were not careful to follow my procedures. The neck of the flask must be heated first to kill the bacteria on the glass; then a heated instrument must be used to break the tip of the flask as it is held high above the head. Immediately thereafter the flask must be sealed again in a flame [Pasteur demonstrates the procedure]. In these difficult researches, while I sternly object to frivolous contradictions, I feel nothing but esteem and gratitude towards those who warn me if I should be in error.

I then devised a conclusive experiment. I boiled a nutritious infusion in a flask with a long curved neck like this one. [Pasteur holds up such a flask.] The tip of the neck was not sealed but left open to the outside air. Thus, there was no hindrance to the entrance of fresh air with its "vital force" as claimed by the advocates of spontaneous generation. But bacteria in the entering air would be trapped by the walls of the long glass tube. The fluid remained sterile so long as the flask was maintained in the vertical position. If, however, I contam-

FIG. 49. *I contaminate the broth with bacteria in the neck of the flask.*

inated the broth by allowing some of it to flow into the neck
[fig. 49] and then back into the flask, putrefaction promptly
followed. So we see that life does not arise spontaneously; life
comes only from life be it a bacterium, a yeast, a tree, a
butterfly, or a baby.

My studies on the diseases of wine and beer and on the
origin of life led me to wonder about the cause of diseases in
animals. About that time–this was in 1865–the important
silkworm industry in France was being destroyed by a disease
that affected the eggs, larvae, and adult moths. The chief
symptom was black spots on the worms which looked like
pepper or *pébré*–so we called the disease pébrine. The Minister
of Agriculture asked me to investigate, but I replied that I was
a chemist, not a zoologist. He insisted. So for love of my
country, I accepted the challenge. I shall not tell you about all

of my frustrations and failures, my disappointments and almost despair. But by perserverance and using the same rigorous methods I had employed in my earlier researches, truth finally emerged. I proved that there were actually two diseases of silkworms: pébrine, caused by small round corpuscles which invaded every organ of the worm, and flachery, caused by an intestinal bacterium. Both diseases were conquered by obtaining healthy eggs from Japan, by beginning new cultures of the silkworm in sterilized containers, and by feeding the larvae with mulberry leaves free from infection. Science again had brought man closer to God. But science cannot advance by the grace of God. *Parbleu!* Science needs money.

I was now convinced that human diseases were also caused by microorganisms. To prove this germ theory of human disease, and thereby serve all humanity, I needed a laboratory, to replace my attic room, and facilities, and assistance. The memory of my beloved mother and father who sacrificed so much for me and of my dear daughters, Jeanne, Cécile, and Camille, who died as sweet children of typhoid fever, made me bold to send a letter to the emperor, Napoleon III, requesting his help. I said: "The time has come when experimental science should be free from its bonds. I need a spacious laboratory in which to conduct medical studies with a new ardor, unrestrained by the insufficiency of material means." I awaited anxiously. Napoleon granted my petition. But, as I sadly learned, there were holes in the emperor's purse through which francs escaped to other less worthy projects promoted by clever politicians. But patience and persistence and the help of influential friends finally triumphed, and my laboratory was constructed. Some of my illustrious predecessors were not so favored. For example, Claude Bernard, whom I admired and loved, had for a laboratory a wretched cellar, low and damp. Perhaps it was there that he contracted the disease of which he died.

One by one the diseases of man have been shown to be caused by microorganisms. [Pasteur signals to the projectionist. "The lantern, Monsieur Baylis, *s'il vous plaît.*"] Tuberculosis is caused by a short rodlike bacterium [fig. 50], as shown by the bacteriologist Monsieur Robert Koch of Berlin. Here is a German whom I can admire and respect. Boils are infections with bacteria called staphylococci, which look like clusters of grapes. Pneumonia is due to diplococci, which are short chains, each of two spherical bodies. Blood poisoning and scarlet fever result from an infection with long chains of cocci, called streptococci. Cholera, from which one of my most valued assistants died recently, is caused by short bent bacteria which we call vibrios, and relapsing fever by spiralled forms, the spirochaetes.

To prove that a disease, animal or human, is caused by a microorganism, one must accomplish the following four steps, called Koch's postulates and named after Monsieur Koch. (1) A microbe must be demonstrated in the body or

FIG. 50. *Causative organisms of diseases.*

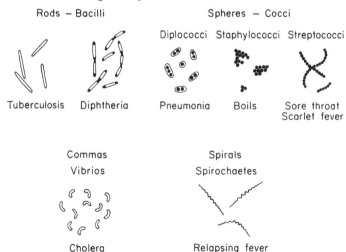

secretions of an animal or person suffering from a particular disease. (2) The organism must be successfully isolated and cultured outside the body in a nutrient broth or on a solid medium. (3) Inoculation of a small drop of the culture into a healthy animal must produce the *same* disease. And (4) the *same* microorganism must be demonstrated in and reisolated from the animal in which the disease was experimentally produced. [To the projectionist: "*Merci beaucoup*, Monsieur Baylis."]

Young men and women, have confidence in a powerful and safe methodology. Whatever your careers may be, do not let yourselves be discouraged by failures, or by any unfriendliness around you, or by the sadness of certain hours which pass over nations. And do not let yourselves become tainted by a deprecating and barren scepticism. Say to yourselves: "What have I done for my instruction?" As you gradually advance in knowledge and understanding, then ask: "What am I doing for my country? And what have I done for the good of humanity?" But whether our efforts are or are not favored by life, let us be able to say as sunset approaches, "I have done what I could."

Fortunately, the body has a marvelous ability to develop a defense against bacteria and viruses, if conditions are proper. We call this immunity. As I discovered in my study of anthrax in sheep, we can immunize an animal by the inoculation of a vaccine, consisting, in that instance, of weakened anthrax bacteria which have lost their power to cause the disease but not the power to stimulate the defenses of the body. But would this method be successful in men, women, and children? Anthrax is not a human disease, but another animal disease —hydrophobia or rabies—can be a terrible human disease. As a child I saw several people die in agony after being bitten by a mad wolf. So I began studying hydrophobia in dogs. I finally succeeded in developing a vaccine from the brains of infected rabbits which would immunize dogs against hydrophobia. Would this vaccine save the life of a child bitten by a mad dog or wolf? I tell a story.

One summer day, not long ago, a lady brought her nine-year-old son, Joseph Meister, to me. He had been bitten by a mad dog in a small town in France's lost province of Alsace, which with Lorraine was taken from us by Bismarck. The lad could hardly walk. There were fourteen bites, some very deep, on his arms and legs. It was then two days later. This boy would surely die in agony. His mother was distraught, and so was I. What could I do? What should I do? I was not a physician, and the doctors in Paris were already critical of this chemist engaged in medical research. If I gave young Joseph the series of inoculations of rabbit brains with attenuated virus and they failed, I could be tried. Maybe even *la guillotine!* *C'est possible.*

I arranged for the care of the boy and his mother, then consulted two physician friends. We considered all aspects, the dangers and the risks. Finally we decided to give the inoculations, beginning with the weakest virus and increasing its strength each day. The twelfth injection was rabbit brain with fully potent virus. That night I could not sleep. But no symptoms developed and Joseph was saved from a cruel death—and so have many people since. Joseph Meister still writes to me, and I to him. I encourage him to study well and to love France.

But biologists should be concerned with more than microorganisms, the germ theory of disease, immunity, and the origin of life. They are also members of society. They have a social responsibility—all scientists—more so than other citizens because of their education and discipline in solving problems. And there is much work to be done. The great goals of the French revolution! They have not been reached in France or anywhere in the world. Men, women, and children are still in prison and in bondage–slaves of other men. Poverty and ignorance still abound. *Liberté* for all? Not yet. In the recent terrible war in America, North was against South, state against state, brother against brother. It was a struggle for the survival of a new nation which President Lincoln has said was conceived in liberty and dedicated to the proposition that all people were created equal. *Egalité?* Not yet. And there is still racial and class warfare and violence, although all of us belong to one species, *Homo sapiens*, and stand as brothers and sisters before our Creator. *Fraternité?* Not yet. Wherefore–as I said to the Académie des Sciences recently–it is the grave responsibility of all scientists (and, I add, of all students studying science) to exercise a concern for social problems and to contribute out of our scientific disciplines to their solution for the sake of freedom and peace, of justice and mercy, and the brotherhood of all mankind. *Mesdames et Messieurs, je vous remercie.* [Pasteur acknowledges the response of the audience with a bow, picks up his cane, and leaves the lecture room on the arm of the assistant, Monsieur Baylis.]

C. D.

Charles Darwin

1809–1882

I PLAY Darwin at about seventy years of age, three years before his death. His delightful *Autobiography* and the reminiscences of his son Francis have guided me in my characterization. Phrases and entire sentences from the *Autobiography* have been quoted, although these passages are not marked. In my live impersonation of Darwin I take advantage of anachronisms, such as a public address system [fig. 51] and the projection of slides, even Kodachrome photographs of the Galápagos kindly given to me by Dr. Robert I. Bowman of San Francisco State University. I am grateful to Dr. Bowman also for a critical reading of this lecture, especially the part on the Galápagos. And my colleague Michael Ghiselin, a leading Darwin scholar, has been very helpful. For the sake of instruction of beginning students in biology on the major points of a modern theory of natural selection, I have "Darwin" place greater emphasis upon two important arguments than that actually presented in his writings. They are survival of an individual organism to the age of successful reproduction and the role of isolation, by one

FIG. 51.   *"Darwin" lecturing to Zoology 10.*

means or another, in the differentiation of a new species. The final scene in which "Darwin" reflects upon an expedition to the moon while observing that "orbéd maiden" from his garden and upon a night on the *Beagle* as it sails the Pacific is a figment of my imagination.

The reading of a recent article by Stephen G. Brush (*Science* 183:1164–1172, 1974) entitled "Should the History of Science Be Rated X?" has caused me to reflect upon this series of lectures in the light of his penetrating comments. Have I unduly "fictionalized" my characters in some instances (Darwin, Mendel, Spemann)? And in other instances (Harvey, Beaumont) have I allowed my "guest" lecturers to overstress the importance of rigorous objectivity? Many scientists and historians of science would object to the former, philosophers and humanists to the latter. Hopefully, I have not failed my students and the readers of these lectures in attempting to soften the image of the "robot-like scientist lacking emotions and moral values" (Professor Brush's phrase), and at the same time in striving to support the traditional image of the scientist as a rigorous "neutral factfinder." But, 'hush, here comes "Charles Darwin."

The Darwin Lecture

[Darwin enters the auditorium clad in a long black cape. He is carrying a large stack of books, which he unloads on the lecture table. He removes the cape, revealing a brown tweed frock coat of Victorian style. His beard is full and white. He moves to the podium where he picks up a microphone and examines it quizzically. He speaks slowly in a sombre voice.]

Ladies and gentlemen. [Darwin looks again at the microphone.] I have been informed that if I speak into this tube my voice will appear to come from heaven. I dislike newfangled contraptions, and I no longer believe in heaven, much to the distress of my good wife, Emma. She is a Wedgwood, you know, and the Wedgwoods are noted for two things: their piety and their crockery.

It is apparent that I have written a good deal [he gestures to the pile of books], and I did not bring all of my books today [fig. 52]. Absent, for example, is my two volume treatise on barnacles. The book which I most enjoyed writing was my first one [he picks it up]: *The Journal of the Voyage of the Beagle*, an account of my round-the-world trip on the H.M.S. *Beagle* when I was a young man of twenty-two years of age to begin with and twenty-seven at the conclusion of the five-year voyage. This trip altered the course of my life and profoundly influenced most of my subsequent work and thought. It was on the *Beagle* that I gained my first insight into organic evolution.

Let me first explain the state of our understanding of this subject in 1831, before the voyage. Species were regarded as fixed and immutable; they had been established by acts of special creation. This was the prevailing doctrine which I accepted without question, and I was not unschooled in biology. I had studied medicine for two years at Edinburgh University in accordance with my father's wish, he being a physician. But I found lectures intolerably dull—maybe this is your experience also—and I had no stomach for human blood. So with my father's permission I gave up this pursuit and chose to become a clergyman. Although I did not understand theology and objected to many of the dogmas of the Church, I thought that I might live a useful life as a country vicar. Accordingly I entered Cambridge University to study the classics. During the three years there I became acquainted with the Reverend Professor Henslow, a biologist, and next to shooting my favorite pursuit became collecting beetles. In fact, I discovered that I had a passion for collecting. To give proof of my zeal, I relate the following incident, which I remember vividly. One day on stripping some bark from a dead tree, I saw two rare beetles. I seized one in each hand. At that moment I spied a third, a new kind, which I could not bear to lose, so I popped the one I held with my right hand into my mouth. [Darwin pantomimes.] Alas, it ejected some intensely acrid fluid which burnt my tongue so that I was forced to spit it out [he mimics the act], which beetle was lost, as well as the third one.

In the summer following my third year at Cambridge, while I was collecting beetles and reading and shooting, I received a letter from Professor Henslow informing me that

FIG. 52.   *I did not bring all of my books.*

Captain Fitz Roy of the H.M.S. *Beagle* was seeking a volunteer to accompany him, without pay, as a naturalist on a voyage around the world. This was for the Royal Navy. I was instantly eager to accept, but my father objected. Fortunately my uncle Josiah Wedgwood came to my defense and persuaded my father to withdraw his objections. And so I boarded the *Beagle* in December with a few books, including the first volume of Lyell's *Principles of Geology* and Milton's *Paradise Lost*, some instruments, a passion for collecting as I have explained, and a

FIG. 53. *She was a ten-gun brig.*

weak stomach which objected to the sea from the moment we left the breakwater until we returned nearly five years later.

[Darwin motions to the projectionist for the lantern slides to be shown.] The first picture (fig. 53) is a sketch of the *Beagle* lying at anchor in the Strait of Magellan. She was a ten-gun brig. Our course is shown in the next picture [a map of the world, fig. 54]. We sailed south and west to South America. [Darwin follows the course of the *Beagle* with a pointer.] You can see that we had many ports of call, affording abundant opportunities for observing and collecting plants and animals and for geologizing. It was in South America that I gave up my gun, preferring to observe rather than to shoot. Since that time I have nurtured an increasing tender concern for all living organisms from the lowly earthworm to *Homo sapiens*. While sailing from port to port I assiduously studied my collections, carried out dissections, made voluminous notes, and read. I was greatly impressed with the magnificent tropical vegetation, the broad deserts of Patagonia and the forest-clad mountains of Tierra del Fuego. The sight of a naked savage in his native land was one never to be forgotten. Likewise unforgettable was the shocking inhumanity of man to man which I saw on haciendas in Brazil. Bound, cringing slaves were beaten unmercifully, and families were torn asunder by the separate sales of husband, wife and children. And again in Argentina the torture and massacre of the Indians was unbelievable.

The *Beagle* sailed around the Horn and up the coast of Chile, making stops for varying lengths of time. While the ship was in port and once when it was beached for repairs, I explored the interior of the country in search of fossils and new species of plants and animals. Considering the infirmities of my later life, I reflect with nostalgia on my vigor and physical fitness as a young man: how I was able to work in the cold and rarefied atmosphere of the Cordilleras when the mules and my Chilean guides were exhausted, or again, back on the pampas, how one day I trudged for eleven hours without water in the blistering desert heat in search of a freshwater source which had been marked on an old Spanish map.

The *Beagle* worked its way up the coast to Peru and then headed west into the open Pacific in September, 1835—almost

FIG. 54.   *Voyage of the Beagle around the world.*

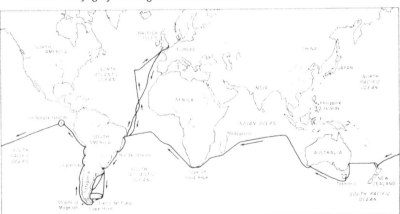

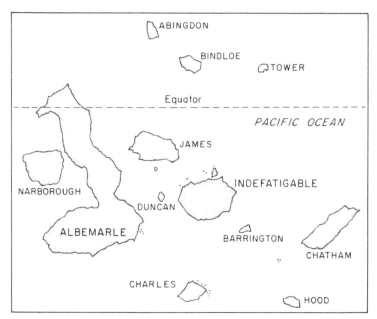

FIG. 55.   *Map of the Galápagos.*

four years after leaving home. Within a week we arrived at the Galápagos archipelago lying on the equator about 600 miles from Ecuador, to which country the islands belong. It was here that I gathered evidence of the mutability of species, to which subject I shall return presently.

From Galápagos the *Beagle* crossed the Pacific stopping at various islands. While in South America I had arrived at a theory on the origin of coral reefs without ever having seen one. It was a great satisfaction to me, therefore, to confirm my theory by observations in the Pacific. Later I wrote a book on coral reefs, which I brought with me today. In it I set forth the evidence supporting my hypothesis that atolls, barrier reefs, and fringing reefs were formed by the action of corals and the elevation or subsidence of the land. My ideas were later accepted by Lyell, and they are now generally approved. The *Beagle* continued to New Zealand, Australia, across the Indian Ocean by-passing Madagascar, round the Cape of Good Hope, and back to Brazil for a stop before returning home by October 1836. My father greeted me with the remark to my sisters, "Why, even the shape of his head has been altered."

Now let us go back to the Galápagos. The archipelago is a group of islands, as we see in the following picture, clustered about the equator [fig. 55]. They are oceanic islands. That is, they have arisen, like Aphrodite, out of the sea, by volcanic action in this instance. Many other islands are continental islands—such as the British Isles, the West and East Indies, Madagascar, et cetera. They were formed by the subsidence[1] of bridges so that the islands with their faunas and floras became separated from the mainland. A continental island is fully populated at the time of origin. Oceanic islands, on the other hand, are at birth sterile of land-dwelling forms. Nature conducts an experiment. She forms an oceanic island and then waits with interest to see who comes to live there. It becomes populated with only forms which can reach it on wings (strong flying birds) or by winds (seeds and spores of plants, insects, and small birds) or by drifting land rafts (reptiles and rodents). One is immediately struck by the total absence of amphibians on oceanic islands, unless introduced by man. Why, one might ask, would the Creator not place toads and frogs and salamanders on oceanic islands when He has so generously distributed them elsewhere in the world, if species arose by special creation? The answer is: amphibians are absent from oceanic islands because they and their eggs cannot survive a voyage at sea. So the Galápagos Islands are sparsely populated in terms of diversity of types of land-dwelling animals.

There is an ugly land iguana and a marine iguana which I call the imp of darkness, because it is almost as black as the lava outcroppings upon which it basks. The marine iguana is an

---

[1] Nowadays the gaps of oceanic waters between some continental islands and the mainland are explained by "drift."

excellent swimmer and diver; and it can remain under water for long periods of time. A seaman on board sank one, with a heavy weight attached to it, thinking thus to kill it; but when, an hour afterwards, he drew up the line, it was quite active. The iguana feeds on seaweeds obtained from the bottom of the coastal water. The giant hissing tortoises are impressive creatures. Incidentally, *galápagos* is the Spanish word for these tortoises. Each island possesses its unique species of the reptile. The inhabitants say that they can distinguish the tortoises from the different islands. I never dreampt that islands about fifty or sixty miles apart, and most of them in sight of one another, formed of precisely the same rocks, placed under a quite similar climate, rising to a nearly equal height, would have been differently tenanted, but this is the case. [The animals mentioned are illustrated with slides.]

And there was a most singular goup of thirteen species of finches to which I gave much attention. They are shown in the next pictures. [The slide, an anachronism, is a reproduction of a figure from David Lack's *Darwin's Finches* which illustrates fourteen species, one more than that reported by Darwin.[2]] As you see, they vary in size from large, heavy-billed seed-eaters, like European hawfinches, to small slender-billed, insectivorous birds like warblers [fig. 56]. When I studied the distribution of the finches on the several Galápagos islands, I discovered that each island had its characteristic finch fauna. Rarely were there more than three species present on any island, and then only on

[2]According to Dr. Robert I. Bowman: "This slide depicts the 13 species of Galápagos finches, as currently recognized, plus one only species of Cocos Island Finch, *Pinaroloxia inornata*. Collectively, the birds of Galápagos and Cocos are popularly known as 'Darwin's Finches.' The 13 species referred to by Darwin are presently considered 8 or 9 under modern taxonomy. Darwin was unaware of the Mangrove finch and the 'Woodpecker' finch, among others. Luckily, however, this number and present numbers agree closely, so the problem for your lecture is somewhat resolved."

Heavy billed; seed-eater   Slender billed; insectivorous
(*Geospiza magnirostris*)   (*Certhidia olivacea*)

FIG. 56.   *Galápagos finches.*

the larger central islands where there was a greater diversification of habitats. It soon became apparent to me that the first
colonizers, probably blown to sea in a storm, came from the
South American mainland, where the nearest relatives of the
Galápagos finches live today. From this ancestral stock the
finches radiated to all of the islands. In time accumulations of
variations led to a diversification among the birds. Those on
Hood Island, for example, became different from those on
Tower Island to the north. Meanwhile, anatomical and phys-
iological changes were occurring in the finches of the other
islands. Later some birds from Hood or Tower might be blown
by a high wind to a third island (Indefatigable, for example) to
compete with the finches of that island. If the newcomers were
preadapted to a habitat on Indefatigable, they might survive
and continue to evolve side by side with the earlier immigrants
to Indefatigable. The divergences between the two popula-
tions were now so great that they did not interbreed. Indeed,
they had become distinct species.

   This adaptive radiation did not and could not occur on
the South American mainland because all habitats were occupied by other kinds of birds which presumably prevented the
evolution of the ancestral, ground-dwelling, seed-eating finch
into species like those on the Galápagos where other kinds of
birds did not exist. So where there was once one species on the

islands, now there are many. And here, both in space and time, we seem to be brought somewhat near to that great mystery of mysteries: the first appearance of new beings on this earth.

My most important book is, of course, *The Orgin of Species*. [Darwin selects it from the stack, fig. 57] It is often times simply referred to as *The Origin*. Actually it has a very long title, namely, *On The Origin of Species by Means of Natural Selection, Or the Preservation of Favoured Races in The Struggle for Life*. After returning to England it appeared to me that I might be able to throw some light upon the origin of species by collecting all the facts bearing on variation of animals and plants under domestication and in nature. I opened my first notebook in 1837. The following year I happened to read a book on population by the theologian Thomas Malthus. Being well prepared to appreciate the struggle for existence which I had observed in plants and animals, I was at once struck with the idea that under these circumstances favorable variations would be preserved and unfavorable ones eliminated. In 1842 I allowed myself the satisfaction of writing a brief abstract of my theory of natural selection. This was expanded in 1844 to 230 pages. My friends, especially Lyell, whom I have already mentioned, advised me to write at length on the subject. So I began on a scale four or five times as extensive as that which later appeared in *The Origin*.

My work on this project was interrupted by other studies, one of them being my two-volume monograph on barnacles, to which I alluded earlier. Actually, I had intended to describe only a single abnormal barnacle from the shores of South America, but I was led, for the sake of comparison, to examine as many genera of barnacles as I could procure. Before long the house was filled with collections of barnacles to which I devoted great attention, so much so that one of my children once asked a neighborhood friend: "And what does your father do with his barnacles?"

FIG. 57.   *My most important book is* The Origin of Species.

My plans for a multivolume publication on the origin of species were overthrown in 1858 by my friend, Alfred Wallace, who sent to me a manuscript intended for publication. Here, to my dismay, was a brief exposition of precisely the same theory, natural selection, on which I had been working for more than twenty years. Of course, I forwarded Wallace's paper to Lyell for publication in the *Proceedings of the Linnean Society*. But Lyell, and other friends, insisted that I also publish an essay in the same issue of that journal. This was done. And so in 1858 the theory of natural selection was launched, but with no great splash in the scientific world. Immediately thereafter my friends strongly advised me to prepare a volume on the transmutation of species. I set to work and although hampered by ill health I finished the task in thirteen months. *The Origin* was published in November of 1859. All copies were sold out on the day of publication. Several editions have followed, and the book has been translated into almost every European tongue.

The success of *The Origin* may be attributed to the fact that I was forced to select the most striking facts and conclusions from a much larger work; that I gave a full and sound documentation of each conclusion; and that owing to many years of thought, I had been able to anticipate and to deal with all serious objections to my views. I had been led to an unshakable belief in evolution by comparing deductions from this hypothesis with those from the alternative hypothesis of special creation. Both sets of deductions were tested by the large body of facts available regarding variation in plants and animals. It was clear that evolution alone agreed with the facts. I then formulated certain hypotheses to explain the cause and the course of evolution. Deductions from these hypotheses were likewise tested against the facts of nature. Natural selection fitted them, whereas alternative hypotheses did not agree with them and had to be rejected.

The essential points of the theory of natural selection for your notebooks are as follows: (1) Variation is an innate characteristic of all plants and animals. (2) More organisms of each kind are born, hatch, or germinate than can possibly survive to reproduce. (3) There is a consequent struggle for existence and for reproduction. (4) Those variations which better enable an organism to survive and reproduce in a given environment will favor their possessors over other less well-adapted organisms. (5) "Successful" variations will be transmitted from generation to generation. (6) Isolation of parts of a population will lead, providing they survive, to diversification and to the formation of new species.

[Darwin picks up another book.] This book, *The Power of Movement in Plants*, is particularly dear to me because I was assisted in experiments here reported by my son, Francis. Frank and I were impressed by the bending of seedlings towards the light. On a hunch that the tip of a shoot is involved in the light-sensitivity and differential growth, we cut the tips of shoots of Canary grass or oat at varying distances from the apex. If less than 0.1 of an inch were removed, the power of bending towards the light was not destroyed. If 0.2 of an inch of the tip were removed, however, no bowing of the shoot occurred. We thought that perhaps the result of the latter instance might be due to an injury of the young shoot, so we performed the following controlled experiment. Over the apices of one group of seedlings we placed thin glass tubes painted black, and to the tips of another group of shoots we added transparent glass tubes of the same size and weight. A third group had no caps. All were exposed to light from one side [fig. 58]. The first group of seedlings, with black tubes, continued to grow vertically without bending, even though the lower parts of the stems were illuminated. The second group, with clear tubes, bowed towards the light just as did the third group of shoots without any caps whatsoever. We concluded that some influence must emanate from the tip of the shoot, conceivably a substance, which is released by illumination and which causes the directional growth of the lower regions of the stem. Perhaps this experiment will be the tip-off [an intentional pun over which Darwin chuckles] for future studies on plant growth.

Another botanical work in which I took much pleasure was a study of insectivorous plants. [Darwin picks up his book, *Insectivorous Plants*.] In the summer of 1860, I was surprised by

FIG. 58. *Seedlings with clear caps or with no caps bent towards the light.*

finding how large a number of insects were caught by the leaves of the common sun-dew (*Drosera rotundifolia*) on a heath in Sussex. I had heard that insects were thus caught but knew nothing further on the subject, one that seemed well worthy of investigation. The results have proved highly remarkable. Firstly, the extraordinary sensitivity of gland-bearing leaf hairs or tentacles [fig. 59] to slight pressure and to minute doses of certain nitrogenous fluids is shown by the movements of the hairs. A particle weighing no more than 1/78,000 of a grain will cause a tentacle to move. I found this so incredible that I rode my hobby-horse to tell all my friends about my latest hobby, begging them to excuse the rider's enthusiasm.[3] Secondly, the viscid secretion of the glands has the power to digest unappetizing insects and nitrogenous substances. The secretion glitters in the sun, hence the plant's poetical name, sun-dew. Thirdly,

[3] See letters to J. D. Hooker and C. Lyell in *Life and Letters of Charles Darwin*, edited by Francis Darwin, p. 492 (see References).

striking protoplasmic movements occur within the cells of the tentacles when the glands are excited. Now considering the havoc wrought by most animals upon plants, it is refreshing to discover that some plants have evolved a clever mechanism for turning the tables.

The figures in this book were drawn by my sons George and Francis. I acknowledge my indebtedness to them and to my other children, who have given me much understanding and tender support through many years of scientific work. A more loving family than mine I have not seen. Indeed, every member of the household is devoted, from footman to our cook who has been with us over twenty-five years. And there is usually much work with the house swarming with Darwins and Wedgwoods, especially on weekends. My butler, Parslow, has

FIG. 59.   *Leaf of a sun-dew,* Drosera rotundifolia.

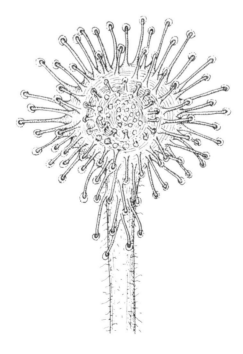

counted on occasions as many as thirty pairs of little shoes to be cleaned of a morning.[4]

And now what of man. Is he, as the psalmist says, "a little lower than the angels and crowned with honor and glory"? Is he a product of special creation, unlike other animals? Is he apart *from* or part *of* nature? As soon as I became convinced that species were mutable, I could not escape the belief that man must come under the same law. Accordingly, I collected notes on this subject for my own satisfaction, with no intent, at first, of publication. In *The Orgin* I did not discuss man but merely stated at the conclusion of the book that my study would throw light on the origin of man and his history. Eventually, I wrote *The Descent of Man* [Darwin picks up this volume from the stack, the full title of which is *The Descent of Man, and Selection in Relation to Sex.*] In this book I marshalled the evidence of the indelible stamp of man's lowly origin. It can be summarized in the statement of my good friend, Thomas Huxley, who more than anyone has defended and championed my ideas. Said Huxley: "Bone for bone, muscle for muscle, blood vessel for blood vessel, man and the apes are alike." I wrote in the concluding chapter of this book that I was aware that the conclusions reached in this work would be denounced by some as highly irreligious. But, I added, he who denounces them is bound to show why it is more irreligious to explain the origin of man as a distinct species by descent from some lower form than to explain the birth of the individual through the laws of reproduction. The birth of both the species and the individual are equally parts of that grand sequence of events which our minds refuse to accept as the result of blind chance.

And this brings me to my ideas on religion. Recall that I once planned to be a vicar. I really never abandoned the plan in

[4]See David Starr Jordan's account of his interview with Darwin's butler in Jordan's *The Days of a Man*, p. 272, World Book Co., Yonkers-on-Hudson, New York, 1922.

a formal way; science simply moved in and took possession of me. Later many doubts grew, and I slowly became an agnostic—one who does not know—to the great concern of my magnificent wife to whom I owe so much. And there were other changes in me which I cannot explain. As a young man poetry gave me much pleasure, especially the works of Milton, Byron, Wordsworth, and Shelley, and I took intense delight in Shakespeare. Pictures afforded considerable enjoyment and music very great delight. But now for many years I cannot endure a line of poetry. I tried lately to read Shakespeare and found it so intolerably dull that it nauseated me. I have also almost lost my appreciation for pictures and music. I retain some taste for fine scenery, but it does not cause me the exquisite delight which it formerly did. This curious and lamentable loss of the higher aesthetic tastes surely results in a loss of happiness, and it may injure one's moral character by enfeebling the emotional part of one's nature. My mind seems to have become a kind of machine for grinding out laws from large collections of facts, but why this should have caused the atrophy of that part of the brain on which the higher tastes depend, I cannot conceive. All this I have recorded in my autobiography, which I recently wrote for my family.

If I have any advice to you it is just this: love science but do not worship it. Put science in its proper place, ranking it along with philosophy and history, music and religion, literature and art. If I had my life to live again, I would make it a rule to read some poetry and listen to some music at least once every week. Perhaps the parts of my brain now atrophied could have been kept active through use.

[Darwin pauses, and then continues as if in a reverie.] I was in my garden recently at dusk, as a full moon rose behind a nearby hill. It was brilliant in the clear evening sky. I wondered how exciting it would be to make a scientific expedition to the moon, to geologize, and to gather information upon its origin and possibly on that of the earth. And as I was reflecting on these scientific matters, there suddenly formed some fleecy clouds which drifted across the face of the moon. In a flash I was aboard the *Beagle* crossing the Pacific and watching the full moon and its silvery path on the shimmering sea as the gentle trade winds filled the sails. And then there came to mind those lines from Shelley's "The Cloud," which I once loved so much.

> That orbéd Maiden with white fire laden,
> Whom mortals call the Moon,
> Glides glimmering o'er my fleece-like floor,
> By the midnight breezes strewn;
> And wherever the beat of her unseen feet,
> Which only–the angels–

[Darwin is suddenly overcome with emotion, fig. 60. He buries his head in his hand, and with a sigh he leaves the podium and auditorium, forgetting his cape and books.]

FIG. 60. *Which only – the angels –*

References

THIS ANNOTATED list of books and articles is not intended to be a bibliography but simply a citation of the sources which I used in preparing the lectures. I have intentionally or inadvertently not included many important works (which scholars in the history of biology may find shocking), but I remind the reader, for the last time, that the lectures were designed for one purpose only: to capture and hold the interest of beginning students in biology.

GENERAL READING

Gabriel, M. L., and S. Fogel, eds.
  1955. *Great Experiments in Biology.* Englewood Cliffs, N. J.: Prentice-Hall. The editors state that the objective of this work is "to convey something of what science is and the way in which science grows." They have selected for republication a connected series of three to ten researches in twelve areas of biology. Each of the dozen sections is introduced by a useful chronology of major discoveries and publications. Most of the chronologies do not include many

works between 1920 and 1955. The last entry in the chronology in genetics, for example, is 1934. Two of my characters, Harvey and Beaumont, were excluded by the editors in the selection of works for reprinting, probably because the book has no section on animal physiology.

Gardner, E.

1972. *History of Biology.* Minneapolis: Burgess. This easy-to-read book includes the works of biologists of our time as well as those of eminent scientists of the past. The reader will find a good chronology of events, a useful bibliography after each chapter, and a listing of general references at the end of the book. Dr. Gardner is a professor of zoology and dean of the School of Graduate Studies at Utah State University, Logan, Utah. He obtained his Ph.D. in zoology at Berkeley under Professor Richard Goldschmidt, to whom both Gardner and I are indebted for stimulation in the history of biology.

Singer, C.

1931. *The Story of Living Things.* New York: Harper and Brothers. This book is an excellent, concise, authoritative thesaurus on the history of biology. Unfortunately, much of modern biology is not included. The substance of this work was delivered as lectures at Berkeley while Professor Singer was on leave from University College, London. I had the good fortune of attending these lectures. The British edition is very similar (*A History of Biology*, Abelard-Schuman, London, 3rd ed., 1959).

Suñer, A. P.

1955. *Classics of Biology.* Translated by Charles M. Stern. New York: Philosophical Library. Many classic papers in biology are reprinted in this book together with commentary by Professor Suñer. The author was a distinguished physician and biologist. He obtained the Chair of Physiology at Seville University at the age of twenty-four. Then he was professor of physiology at Barcelona, recipient of many international honors, and author of numerous books.

WILLIAM HARVEY READINGS

Chauvois, L.

1957. *William Harvey: His Life and Times, His Discoveries, His Methods.* New York: Philosophical Library. This authoritative work is one of several biographies of Harvey which could be cited. It was published on the tricentennial of Harvey's death. As pointed out in the foreword by Dr. Zachary Cope, Dr. Chauvois "carefully read, studied and sifted all available information about Harvey. In addition, as an accomplished scholar, he has very closely studied the Latin texts of Harvey's published writings." The book is readable and gives one the feeling of having known Harvey. I must place myself, however, among those who say, to quote Chauvois, "that the author exaggerates and that the language is too extravagant." But that author adds, "my mind and imagination have been so conquered and subjected by my hero that I would almost deify him."

Harvey, W.

1928. *Exercitatio anatomica de motu cordis et sanguinis in animalibus.* With an English translation and annotations by Chauncey D. Leake. Springfield, Ill.: Charles C. Thomas. This book is a tricentennial facsimile of Harvey's great work in the original Latin text of 1628. Then follows Professor Leake's translation, upon which I commented in the preface to the Harvey lecture. The translation has helpful annotations by Leake as footnotes, and there is a chronology of Harvey's life and times at the end of the book. Dr. Leake was formerly a professor of pharmacology at the University of California Medical Center in San Francisco. I am indebted to him, along with Charles Singer (see above), Charles Kofoid, Herbert Evans, and other distinguished historians of biology who lectured in Zoology 117, our course in the history of biology, which I attended as a student in the thirties.

1907. *An Anatomical Disquisition on the Motion of the Heart and Blood in Animals.* Translated from the Latin by Robert Willis. London: J. M. Dent. In addition to Willis's translation, this book contains an introduction by the editor, E. A. Parkyn, some letters of Harvey's, his last will and testament, and a delightful account, the "Anatomical Examination of the Body of Thomas Parr." The post-mortem examination was conducted by Harvey at the king's command on Thomas Parr, who died at the age of 152. Harvey's observations on pollution, the ill effects of too much food and drink, etc., are priceless. Willis's translation of Harvey's treatises have been standard texts.

Willis, R.

1878. *William Harvey. A History of the Discovery of the Circulation of the*

Blood. London: O. Kegan Paul. This book contains an excellent summary of the antecedents of Harvey's works and thoughts. I know very little about Dr. Robert Willis, except that he was a physician and a scholar of history and philosophy. He wrote a book on urinary diseases published in 1839.

WILLIAM BEAUMONT READINGS

Beaumont, W.

1941. *Experiments and Observations on the Gastric Juice and the Physiology of Digestion.* New York: Peter Smith. This is a facsimile of the original edition of 1833 together with a biographical essay by Sir William Osler entitled "A Pioneer American Physiologist." William Osler probably needs no introduction to many readers. He was Canadian by birth, and he held successive academic or medical posts at McGill University, University of Pennsylvania, Johns Hopkins, and Oxford (Regius Professor of Medicine). A colleague eulogized him as follows: "At the time of his death he was probably the greatest figure in the medical world, the best known, the most influential, the most beloved." Osler's enthusiasm for Beaumont and his work has special meaning.

Myer, J. S.

1912. *Life and Letters of Dr. William Beaumont, including hitherto Unpublished Data concerning the Case of Alexis St. Martin.* St. Louis: C. V. Mosby. This book contains 58 illustrations and a foreword by Sir William Osler, who writes that Beaumont's family papers were placed at the disposal of Dr. Myer, "to whom it has been a labor of love to present a picture of the first great American physiologist." The task of sifting through trunks of papers was formidable. Beaumont had preserved not only all letters to him but a rough draft in his handwriting of those sent by him. After his exhaustive and scholarly research, Dr. Myer states in the preface that he felt like Boswell in preparing the biography of Samuel Johnson. Wrote Boswell: "I was obliged to run half over London in order to fix a date correctly, which, when I accomplished, I well knew could obtain me no praise, though a failure would have been to my discredit." Dr. Myer was on the medical staff of Washington University in St. Louis.

Rosen, G.

1942. *The Reception of William Beaumont's Discovery in Europe.* New York:

Schuman's. This book documents the claim that Beaumont was more rapidly and more fully appreciated abroad than at home. The Germans valued his rigorous science (Beaumont's book was translated into German within a year of its publication), the English were more impressed with the clinical implications, and the French were stimulated to undertake studies involving animal experimentation.

HANS SPEMANN READINGS

Mangold, O.
1953. *Hans Spemann. Ein Meister der Entwicklungsphysiologie. Sein Leben und Sein Werk.* Stuttgart: Wissenschaftliche Verlagsgesellschaft. This review and appreciation of the life and work of Spemann was written by one of his early students and assistants, Otto Mangold, who became Spemann's successor as director of the Zoological Institute in Freiburg upon Spemann's retirement in 1939. Otto Mangold married Hilde Pröscholdt, the graduate student and assistant who shared with Spemann the discovery of the organizer. Mangold's biography is replete with authoritative information, although unbalanced by his great admiration for his subject—a fault of most biographers, including this playwright. I had the opportunity of studying with Otto Mangold in the fall of 1935, before I went to Spemann's laboratory, when Mangold was professor of zoology at the University of Erlangen. I acknowledge my indebtedness to Professor Mangold for his warm hospitality and instruction in certain subjects, not including politics, however.

Spemann, H.
1938. *Embryonic Development and Induction.* New Haven: Yale University Press. This well-illustrated summary of the work which occupied Spemann and his students and assistants during the major part of his life is based upon the Silliman Lectures which he delivered at Yale University in 1933. The German edition of these lectures, entitled *Experimentelle Beiträge zu einer Theorie der Entwicklung,* was published in 1936 by Julius Springer, Berlin. As stated in the preface to the Spemann lecture, I have drawn heavily upon this book.

Spemann, F. W., ed.
1943. *Hans Spemann. Forschung und Leben.* Stuttgart: J. Engelhorns Nachf. Adolf Spemann. Spemann began an account of his life and

work, but it was never completed. The editor, Spemann's eldest son, published in this volume his father's recorded remembrances of the early part of his life up to his first lecture as a *Privatdozent* at the University of Würzburg. The reader will also find in this book other writings (in German) by Spemann including letters and his Nobel Lecture before the Karoline Institute in Stockholm in 1935. The editor has added chapters about other periods of Spemann's life, and Spemann's last assistant, Eckhard Rotmann (one of my best friends in Germany before his untimely death), has written a chapter on Spemann as he was known by his associates at the Zoological Institute in Freiburg.

GREGOR MENDEL READINGS

Iltis, H.
  1932. *Life of Mendel.* Translated by Eden and Cedar Paul. New York: W. W. Norton. Iltis was a countryman of Mendel's. In fact, he was born and educated in Brünn, and at the time of publication of his biography of Mendel (first published in German in 1924) he was a teacher of natural science in the same city.

Mendel, G.
  1866. *Versuche über Pflanzen-Hybriden. Verhandlungen des naturforschenden Vereines in Brünn* 4:3–47. English translation first published in 1901 in the *Journal of the Royal Horticultural Society* 26:1–32.

  1869. *Über einige aus künstilicher Befruchtung gewonnenen Hieracium-Bastarde. Verhandlungen des naturforschenden Vereines in Brünn,* 8:26–31. English Translation first published in W. Bateson, *Mendel's Principles of Heredity. A Defence.* Cambridge University Press, 1902.

Stern, C., and E. R. Sherwood
  1966. *The Origin of Genetics. A Mendel Source Book.* San Francisco: W. H. Freeman. I have referred to this book in the preface to the Mendel lecture. In addition to a foreword by Curt Stern and Eva Sherwood's translations of Mendel's papers on hybrids in peas (*Pisum*) and hawkweeds (*Hieracium*), the book contains translations of Mendel's letters to Carl Nägeli and the papers by Hugo de Vries and Carl Correns in which Mendel's work was "rediscovered," plus letters of de Vries and Correns on how they were led to Mendel's papers. Finally, there are two penetrating discussions by the dis-

tinguished statisticians R. A. Fisher and Sewall Wright on the reliability of Mendel's data and the possibility that he subconsciously allowed bias to influence the counting of his peas.

LOUIS PASTEUR READINGS

Delaunay, A., and H. E. Kouby.

1972. *Institut Pasteur.* A special bulletin issued by the Pasteur Institute. It contains brief illustrated sections on the museum, library, and chapel. One of the figures is a reproduction of the painting of Pasteur which hangs in the museum. It is a well-known portrait of Pasteur with his young daughter Camille, who died at the age of two years of typhoid fever. The impressive figure of Pasteur was the basis of my make-up. I am indebted to Dr. Armelle Kernéis of Paris for a copy of the bulletin.

Dubos, R. J.

1950. *Louis Pasteur. Free Lance of Science.* Boston: Little, Brown. Among the several biographies of Pasteur, this one is highly regarded. It contains a chronology of Pasteur's life and a good bibliography.

Nicolle, J.

1961. *Louis Pasteur. A Master of Scientific Enquiry.* London: Hutchinson. This biography is short and concise. It too has a chronology of events in the life of Pasteur and an annotated index of names.

Porter, J. R.

1972. "Louis Pasteur Sesquicentennial (1822–1972)," *Science* 178:1249–1254. I have commented on this article in the preface to the Pasteur lecture.

Vallery-Radot, R.

1926. *The Life of Pasteur.* Translated from the French by Mrs. R. L. Devonshire. With an introduction by Sir William Osler. New York: Doubleday, Page. I have commented on this biography in the preface to the Pasteur lecture.

CHARLES DARWIN READINGS

Darwin, C.

Of the many editions of Darwin's works, these are the ones I consulted in preparation for the lecture, listed in order of reference there.

1913. *Journal of Researches into the Natural History & Geology of the Countries Visited during the Voyage Round the World of H.M.S. 'Beagle' under the Command of Captain Fritz Roy, R.N.* London: John Murray.

1896. *The Structure and Distribution of Coral Reefs.* New York: D. Appleton.

1896. *The Origin of Species by Means of Natural Selection; or, The Preservation of Favored Races in the Struggle for Life.* New York: D. Appleton.

1896. *The Power of Movement in Plants.* New York: D. Appleton.

1875. *Insectivorous Plants.* New York: D. Appleton.

1896. *The Descent of Man and Selection in Relation to Sex.* New York: D. Appleton.

1959. "The Autobiography" in *The Life and Letters of Charles Darwin, including an Autobiographical Chapter, Edited by His Son Francis Darwin.* Foreword by George Gaylord Simpson. New York: Basic Books. The frontispiece to volume two of this work is the famous portrait of Darwin in black cape and slouch hat leaning against the trunk of a tree. This figure was the basis of my Darwin make-up.

Ghiselin, M. T.

1969. *The Triumph of the Darwinian Method.* Berkeley and Los Angeles: University of California Press. This recent book by my colleague and leading Darwinian scholar, Michael Ghiselin, is a profound and widely acclaimed work. One might think that after a hundred years and many studies of Charles Darwin that there would be nothing more to say on the subject. Yet several recent works have appeared, Ghiselin's among them. Michael Ruse, who reviewed three of them (see *History of Science* 12:43–58, 1974), has referred to the seemingly unending analysis of the great evolutionist as the Darwin industry. Another review (of *Darwin on Man* by Howard E. Gruber) was published by George Gaylord Simpson (*Science* 186:133–134, 1974). Simpson's review was helpful to me on Darwin's method.

Irvine, W.

1955. *Apes, Angels and Victorians; the Story of Darwin, Huxley, and Evolution.* New York: McGraw-Hill. This is a delightful and scholarly book. Irvine was a professor of Victorian literature at Stanford University.

Moorehead, A.

1969. *Darwin and the Beagle.* New York: Harper and Row. This work is a fascinating account of the voyage of the *Beagle*, beautifully illustrated. The text has many quotations from Darwin's books and letters. Alan Moorehead was born and educated in Melbourne, Australia, was a war correspondent during the Spanish Civil War and the Second World War, and has written several other books.

Addendum at second printing.
Several books on Darwin or his theory of natural selection have appeared since *Great Scientists Speak Again* was published. Among them three are recommended to the general reader.
Gould, S.J.
  1977. *Ever since Darwin: Reflections in Natural History.* New York: W. W. Norton.
  1981. *The Panda's Thumb: More Reflections in Natural History.* New York: W. W. Norton.
Stone, Irving
  1980. *The Origin A Biographical Novel of Charles Darwin.* New York: Doubleday.

SOURCES OF ILLUSTRATIONS

Most of the photographs of the characters have not been published heretofore. The portraits of "Beaumont" and "Mendel" were published in *Bios* (see acknowledgments) and that of "Darwin" in the author's *The Third Eye* (Univ. Calif. Press, 1973). A few of the portraits appeared also in newspaper and magazine articles. Many of the line figures are original, others are based upon illustrations in a book by Spemann (*Embryonic Development and Induction*) and in two books by Darwin (*Voyage of the Beagle* and *Insectivorous Plants*). The following figures were taken from other sources.

Figs. 18-21: based upon figures in Morgan, T. H., *Experimental Embryology*, Columbia University Press, New York, 1929. See Morgan for references to the publications of Roux and Driesch.

Fig. 23: from Spemann, H., *Zeitschr. f. wiss. Zool.*, 132: 105-134, 1928.

Fig. 25: from Schmidt, G. A., *Roux's Archiv*, 129: 1-44, 1933.

Figs. 26, 27: from Mangold, O. and Seidel, F., *Roux's Archiv*, 111: 494-665, 1927.

Fig. 31: from Spemann, H., *Zool. Jahrb. Abt. f. allg. Zool. u. Physiol. d. Tiere*, 32: 1-98, 1912.

Fig. 32: from Mangold, O., *Roux's Archiv*, 117: 586–696, 1929.

Fig. 35: from Holtfreter, J., *Biol. Zentrbl.*, 53: 404–431, 1933.

Fig. 45: based on a photograph of Mendel's microscope in Iltis's *Life of Mendel*.

Fig. 51: The photograph of "Darwin" lecturing to Zoology 10 was taken by my friend and colleague Dr. O. P. Pearson. This is one of many photographs of me in my various academic activities that were chronicled in Pearson's *A Professor's Days* (1971), privately printed.